METAVERSE –

A PEEK UNDER THE HOOD

(COMPREHENSIVE GUIDE)

Copyright 2022

John Buxton Publishing

Introduction

Throughout history, the globe has seen many changes and advancements in several facets of life, including medicine and transportation. As a result, some people see what is happening in the tech industry as a revolution. In contrast, others see it as an ongoing process of growth and expansion into the Metaverse or digital reality.

In any case, a significant event is occurring in the field of information and communication technologies, which may result in a Big Bang and lead to the redoing of the entire planet. It may be funny, or it could be dangerous.

Will a brand-new world come into existence, and will this one eventually consumes the one we live in? Is it conceivable that technological advancement will give rise to a brand-new ecosystem known by its working title, i.e., the Metaverse?

The Metaverse can be constructed based on the evidence provided by today's technology, businesses, user experiences, and social trends. Blockchain, cryptocurrencies, augmented reality and virtual reality, intelligent systems, and the Internet of Things (IoT) are examples of some of the technologies that will be a part of the Metaverse when it is fully developed. The distinction between what we know as the real world and virtual worlds will become increasingly hazy as work is done in the Metaverse, which will make use of the complete gamut of types of technology, including overlaid and mixed reality.

An intriguing glimpse into the universe of the Metaverse is dissected in great detail in this book. In the realm of technology, the Metaverse is a fascinating breakthrough that presents a variety of potential. Please get familiar with this area and investigate the possibilities it offers.

Let's dive in for more information!

CONTENTS

Chapter 1

Understanding the Metaverse

The word combines the English word **"universe"** and the Greek word **"meta**." The term "Metaverse" is frequently used to express the idea of a future version of the Web consisting of connections united by a perceptual virtual synthesis. The prefix "meta" (meaning "transcendence") and the root "verse" (inflection of "universe") make up the phrase. a three-dimensional virtual space that is shared and permanent in the cosmos. To stop email spam, the term "Metaverse" and associated ideas were first presented in the area of computer security.

What exactly is a Metaverse?

A variety of up-and-coming technologies, such as virtual reality, augmented reality, and eye-tracking, are combined to create the Metaverse. As a result, the best possible virtual encounter within a virtual environment is created with all of these components.

The contents of the Metaverse

Many people still do not understand what the Metaverse is made of. However, having personally explored the Metaverse, I have become aware of several of its features.

Various features support a virtual economy in the Metaverse. These include 3D avatars, digital assets, games, enterprises, and numerous events. You can organise business meetings, meet up with pals, participate in virtual events, and even monetise your work as a user. You might quickly discover as you explore the Metaverse that it offers many of the same qualities you might find in the real world.

The main distinction with the Metaverse is that you may visit anywhere in the globe using nothing more than your computer and, if you have any, VR goggles, all from the convenience of your own home. Additionally, you can teleport to various locations. For example, in the Metaverse, you can teleport to the following city or even to another room.

What makes the Metaverse significant?

It makes sense that many people could underestimate the significance of the Metaverse in modern culture. What makes the Metaverse so crucial, then?

The Metaverse is significant because it gives people a physical means of virtual communication and connection from any part of the world. The Metaverse also supports a whole trading system where users may engage in various activities like starting a business, interacting with others, and even holding family events. For example, consider just for a moment that you cannot spend the holidays with your family in person. Instead, you put on your headset and digitally join them in the same space.

You may still converse with one other, play games, and even go to other Metaverse worlds from the comfort of your own home. The same is true of training sessions and business meetings. You, your co-workers, and the presentation may all be seated around the same board and discussed in-depth in an immersive setting. Then, when your meeting is finished, you can take off your headset and return to your family's home.

Furthermore, thanks to the Metaverse, you can establish a genuine company within the boundaries of any virtual property you own. A VR clothes store, for instance - or even monetizing your abilities where you can charge customers a fee in compensation for an architectural form or other digital product are just a few examples of how you might use your skills to make money.

As a result, there will quite clearly be a greater need for developers, virtual architects, and other skills, which will lead to more job possibilities. Your creativity truly is the only constraint, and the Metaverse is no exception. In the end, the Metaverse enables you to realistically be present anywhere, at any time, and with everyone. Your connections and your capacity to obtain practical instruction and direction from home may both benefit from this ease.

Perspective from the past

In his 1992 book "Snow Crash," science-fiction writer Neil Stevenson first used the word "Metaverse." In this 3D virtual reality, which makes use of real-world metaphors, humans serve as avatars and communicate with software agents.

In the bleak future depicted in his 1992 book, the Metaverse is a virtual realm where most people spend their time. The idea of the Metaverse has its roots in the mid-twentieth century, but Neil Stephenson's book was the first to establish it in the public eye. It has an impact on online businesses like Facebook, Amazon, and Google. In addition to the companies introducing the concept of the Metaverse to the public, Snow Crash also had an extremely profound effect on some of Silicon Valley's most well-known pioneers, including Sergey Brin, a co-founder of Google, and the founders of Amazon, Blue Origin, and Facebook.

The Metaverse, according to the Facebook CEO, is an online virtual environment that we may enter and explore utilizing augmented reality and virtual reality viewers. He is aware that the Metaverse is referred to as the most recent development of the Internet, a form of virtual reality where any virtual contact can have an immediate influence on the real world. A permanent virtual area made up of the total of all virtual worlds, augmented reality, and the Internet, the Metaverse is an international project virtual space that integrates physical and augmented virtual reality.

Since the publication of Neil Stephenson's book, improvements in Internet speed and processing power have made it possible to create actual applications based on the idea of the Metaverse. In the 20th and 21st centuries, the first decentralized Metaverse experiments debuted. A full Metaverse, an online virtual environment with augmented reality, virtual reality, 3D holographic portraits, video, and other media, has since been made possible thanks to a number of products.

Given that it frequently appears tied to the physics and electrical circuitry of our reality, the Metaverse itself might resemble the physical universe. It need not, however, be the same as the real world. In another sense, the Metaverse, a three-dimensional place created by Mark Zuckerberg, Neil Stephenson, and others, will link the various virtual worlds that already exist. The idea of Web3, which refers to decentralized Internet services where users have more personal control over the information they share online, is evident in these Metaverse representations as well.

The real Metaverse won't be confined to just one game, virtual environment, software, or business, which is how video conferencing apps differ from reality. According to Meta (previously Facebook), the Metaverse will eventually enable us to interact in social, commercial, and educational settings. Microsoft also appears to be concentrating solely on the virtual office area for the time being. Virtual identities, avatars, and inventory are typically connected to an outlet on platforms. However, this alternate reality might enable us to travel anyplace as quickly as we can transfer a profile image from one social media site to another.

- The Metaverse is a colossal network of continuously evolving 3D environments and simulations that may be experienced simultaneously by an essentially infinite number of users, each of whom has a sense of presence.

- Online game universes like Roblox, Fortnite, and Minecraft already contain hints of the Metaverse. Other fantasy books and films like Ready Player One also include the concept of a

Metaverse. Before the publication of American author Neil Stevenson, virtual worlds with Metaverse-like characteristics may be found in several MMORPGs (massively massive multiplayer role-playing games) and other novels with retro-futuristic themes. Since then, the line separating the virtual and real worlds has become even more hazy thanks to films like Ready Player One, video games like Pokémon Go, and fashions like NFTs. This is the fundamental idea behind the Metaverse.

+ The Vmerse Metaverse, a consumer-facing, real-time virtual reality model created to assist real-world applications like college recruiting, alumni methods utilized, crises, etc., was introduced in 2004 by Indian American inventor Bhargav Shree Prakash. After learning about the Metaverse, you might be curious about what you can accomplish there. Again, the solution is simple: virtually everything you can do in real life is also possible in the Metaverse.

Some interesting examples

+ There are many different games and entertainment options available in the Metaverse. Don't picture a traditional video game, though, as using your avatar will completely immerse you in a three-dimensional universe. Instead, let's imagine that you choose to play a great round of poker. In this scenario, you and your opponents will be seated around a gaming table. You have the chance to either gain, or lose cryptocurrency during the game.

+ You can engage in any physical activity in parallel universes, alone or with a team. You won't have any more excuses after this because any training will be feasible no matter the weather. You'll be able to become in shape in this manner and always have fresh stimuli to boost your outcomes and performance. Additionally, the Metaverse will be a challenge for friends and family.

+ Given the evolution of smart working in many businesses, even if not everyone is convinced yet, you will have numerous new job prospects with the Metaverse. Yes, since you can arrange genuine business meetings between so many people using these virtual realities, you can even set up virtual workspaces where all avatars are present at once. In other words, our alter ego will sit next to our co-worker. Therefore, it is a novel technique to avoid scheduling dull meetings using Meet and Zoom from the comfort of our homes!

+ The Metaverse is an excellent resource for learning and training! Consider all the virtual and muti lessons that may be created to captivate and include pupils of all ages and academic levels.

+ Online shopping in another universe. The importance of online buying has grown significantly. However, it might still advance farther with the Metaverse because there are actual stores in three aspects that are identical to those, we can visit to make any kind of transaction in this virtual reality. As a result, our avatar will be able to visit these establishments, try on clothes and shoes, receive assistance from virtual salespeople, and ultimately make various items that he will pay for using cryptocurrency.

Travel is also conceivable within the Metaverse. How many occasions have you wanted to take a trip to another country but were prevented from doing so because of obligations to your family, job, or financial situation? You can quickly get around all of these restrictions with the Metaverse because you can design your own journey in any way you like. You will feel totally immersed in reality when using your avatar, and you will be able to feel all the feelings and sensations you would have if you were actually on a trip.

Chapter 2

The Metaverse is not Facebook

A new surge of interest in the Metaverse has been inspired by Facebook's recent rebranding & investment in Meta. It's everywhere in the news, in memes, on gaming platforms, and on social media. Our physical lives seem to be about to be consumed by corporate pixels and commercial interactions as a result of the word's increasing ubiquity.

However, the Metaverse has been mentioned in advertisements from Roblox and Fortnite for quite some time, and the phrase itself is actually rather old. In Mark Zuckerberg's version, you attend board meetings as an avatar using Quest VR and utilize a gadget on your wrist to covertly communicate with pals while in meetings. When you are outside, you wear augmented reality smart glasses that record what you see and hear. The Metaverse, where you work, shop, exercise, socialize, watch movies, and play games, will be available through phones, laptops, wearable devices, and/or headsets.

Since the 1960s, people have been working to develop fully realized virtual worlds, with the efforts of the video game and film industries fueling this pursuit. Second Life, an alt-reality video game where you play through an avatar and may do just about anything, including build a house or get married, was founded way back in 2003, and it is clearly one of the most frequently mentioned examples of the Metaverse.

There were sufficient fans around the world by 2006 to convene for a summit. The Metaverse is "an extraordinarily immense network of persistent, real-time generated 3D environments and simulations," according to key individuals in the digital world.

The continuity of identity, artefacts, history, and payments should be upheld in the Metaverse. There is a limitless number of people who can experience it at once, and each will feel their unique feeling of presence. Blockchain technology might be used to pay for products that we can take with us through various experiences in an immersive virtual reality that allows people to be present. Imagine being able to virtually put on your Gucci T-shirt or your Animal Crossing award, for instance.

The Metaverse will, without a shadow of a doubt, continue to grow and develop. This concept has a big impact on Zuckerberg's digital world. Zuckerberg's avatar in his Facebook Connect presentation switched between platforms while still sporting the same black T-shirt to show "continuity of identity and things." Mark Zuckerberg's Metaverse is well on its way to housing an infinite number of people, with almost 3 billion Facebook users. Throughout his presentation, Zuckerberg emphasized numerous times how each Metaverse feature will create a "feeling of presence," or awareness of even virtual existence.

Even in the gaming world, discussions about the Metaverse frequently involve Epic Games' Roblox and Fortnite, which some claim are much closer to reality than Zuckerberg's Meta. Both video games satisfy the requirements for persistent virtual worlds. Each has a sizable user base that congregates to play and interact in areas where things (clothes and skins) and money are persistent (Robux and V-Bucks). Millions of people attended the Ariana Grande concert in Fortnite, and together with the ability to customize avatars and emotes, these events foster a sense of presence. The Metaverse is inaccurate, and that is the most crucial fact to understand.

The Metaverse is a goal for many investors, engineers, scholars, and futurists, and Zuckerberg has made it quite apparent that it is a goal for him. However, the public has not embraced Zuckerberg's concept and has little faith in the possibilities of the Metaverse of Meta, an idea that is both intriguing and terrifying to others. A symposium called Welcome to the Metaverse was organized by Rhizome, a nonprofit arts organization that oversees initiatives to archive digital art and culture.

In contrast to many of the democratization expectations we originally had for the Internet, artist David Rudnick has stated that "the notion of the Metaverse is the ultimate centralization" in the passage. Fundamentally, the worries and apprehensions about the Metaverse have grown. Any growth of the virtual world is likely to accentuate its worst characteristics. What would it imply if a small number of for-profit businesses mediated so many crucial interactions? It doesn't give off much hope, if Meta's present dominance on social media is anything to go by. The Metaverse is currently largely conjecture, fantasy, and the hopes of a select few, with numerous gaps where the chilly winds of the unknown blow.

Chapter 3

Characteristics of the Metaverse

Speaking of the fundamentals, the following is a list of some characteristics that characterise a Metaverse:

BOUNDLESS

The Metaverse removes all kinds of barriers, both physical and psychological because it is a 3D virtual environment. There are simply no restrictions whatsoever on the number of individuals who can use it concurrently, the different kinds of activities that can be carried out, the industries that can operate there, etc. As a result, more people can access it than on current Internet platforms.

PERSISTENT

You can't unplug, restart, or reset a Metaverse. Instead, users can quickly and easily enter the Metaverse at any time and from any location, and their experience will remain consistent. Based on the collective contributions of its users, such as the content and experiences they create, a Metaverse will develop over time.

DECENTRALIZED

The users of the Metaverse, who can also take ownership of their personal data, own it instead of a company or a single platform. This is primarily made possible by blockchain technology (more on this later), ensuring that all transactions in a virtual environment are transparent, traceable, and secure always.

IMMERSIVE

You'll be able to access a new level of immersion and engagement, where all human senses are wholly engaged. As a result, people feel more present in their experiences, whether using a VR headset, AR glasses, or smartphone. In addition, the Metaverse, a highly realistic environment, will be able to adjust to its users since they have direct control over things like its environments, items, colours, lighting, and more.

ECONOMIES - VIRTUAL

Participants in the Metaverse can participate in decentralised virtual economies supported by cryptocurrencies, such as the SENSO of Sensorium Galaxy. This includes online markets where users can trade digital goods like avatars, virtual clothes, NFTs, and event tickets for other goods.

SOCIAL ENCOUNTERS

Its users are the Metaverse's living, breathing heart. Every virtual world user participates in co-experiences and contributes to co-creating the Metaverse's future through user-generated content, such as avatar interactions with AI-driven avatars and virtual creations of their own. A real Metaverse is an ongoing experience that combines components from several platforms and target audiences.

The Metaverse's Layers

You will discover a little more about the seven realms of the Metaverse in this section of the guide. The Metaverse can be thought of as having levels that enclose its users. People need to remember that these layers are not in any specific order. However, as individuals explore or improve these levels, they become more interconnected with the Metaverse. As they become more present or more powerful in each layer, they become more fully residents of the Metaverse.

An encounter with a non-materialized reality

The Metaverse is frequently perceived as a 3D environment. And that's indeed how most users engage with the Metaverse. VR headsets, for instance, imitate 3D surroundings. Thanks to interface devices, users can grip or move objects in those 3D environments. The Metaverse, however, isn't a 3D environment. It's also not a 2D space.

In contrast, the Metaverse is a dematerialized reality in which spatial dimensions are meaningless. For instance, Alexa can be found inside small items connected to enormous data volumes. Additionally, individuals within the Metaverse have access to substantial virtual territories that ultimately exist as something as small as a hard disc.

The Metaverse essentially destroys people's perceptions of spatial dimensions. And it's not only the Metaverse itself that's like this. Just as humans can find components of the real world inside the Metaverse, the Internet of Things (IoT) can bring the Metaverse into the physical world.

Exploration and Discovery of a vast and living world

Exploration of the Metaverse involves more than just traveling through a virtual 3D scene. The opportunity to be among one of the first to discover and experience brand-new sights in the Metaverse's constantly expanding landscapes appeals to many individuals.

However, as was already mentioned, the Metaverse is more than just a mass of three-dimensional spaces. Inbound discoveries are also clearly a big part of the exploration and discovery process. For instance, users will continually access content made by the community. Additionally, businesses are often on the lookout for individuals with hobbies and complementary ones.

The Metaverse also enables outbound discovery. (For instance, what most people consider as adverts or spam would be regarded as outbound discovery). But there is a characteristic shared in all these different situations. They focus on individuals being able to discover fresh experiences in the Metaverse.

A thriving economy for artists

The Metaverse is home to an extremely robust economy. As we have already said – cryptocurrency based blockchain based exchanges are commonly used in Metaverse. however, the economy of the Metaverse involves a heck of a lot more than just handing over money for various products and/or services. The Metaverse's economy is totally distinct from the real world because it is what is known as a 'creator' economy.

The early web and the Metaverse have both undergone two distinct cycles of growth. The pioneer era is the first cycle. To produce any material in the digital sphere during the pioneer phase, one needed to be highly skilled. To produce material for the early Internet or Metaverse, one must be an experienced coder. The engineering era followed, where new technologies made it simple for even inexperienced coders to produce content.

The creator era is what distinguishes a true Metaverse, though. Here, everyday people can produce items. They can also sell items with ease.

Spatial computing blurs the line between the real world and the virtual world

As was already mentioned, the boundaries of spatial dimensions frequently disappear in the Metaverse. The lines that are between the analogue and the digital worlds are more blurred as the Metaverse grows. What is real and what is virtual? The idea that the digital world isn't real is simple to state. But what if your money, unique purchases, and even real estate are all done online? That makes it feel authentic. Similarly, a VR space can have little to do with the substance of that place.

In the meanwhile, 3D spaces can be simulated in digital worlds. You can stroll along a virtual trail and observe tangible results. You can travel virtually long distances, burn calories by moving your arms or legs, and simply unwind while gazing out into far-off horizons. In some ways, you aren't moving very far when judged in analogue terms.

You can be moving slowly or along a short track. But the emotion is what counts.

An interoperable experience with decentralized components

A single entity or single component manages a completely digital system in centralized computing. The simplest approach to utilize or construct networks is in this manner. However, it doesn't give end users a lot of autonomy or authority. In contrast, the Metaverse depends on decentralized evolution. This indicates that there are several separately owned and made components in the Metaverse.

Decentralization's precise definition differs depending on how it is applied. Interoperability, however, is a fundamental concept that all facets of the Metaverse share. Based on standards, people produce components that are compatible with one another. This implies that parts can effectively be taken out and replaced. This is comparable to taking out the RAM from a computer and replacing it with a RAM stick made by a different company. The blockchain and other systems are compatible with the Metaverse thanks to shared standards. It also implies that individuals can create their own extensions or applications.

Direct interaction-enabling human interfaces

All so often, people are unaware of how frequently they use technology. For instance, a smartphone is less of a phone and more of a powerful networked supercomputer. To further facilitate engagement, these devices also include a variety of sensors and even a small amount of AI. The majority of people engage with it throughout the day without giving it much thought.

Smartphones aren't just becoming more compact and user-friendly. Additionally, they are being incorporated into Metaverse interface technology. As an illustration, the Oculus Quest essentially combines VR and cell phones. A human interface into the Metaverse is successfully created through the integration of components. In a significant part, people are evolving into cyborgs. Everyday home appliances are becoming smarter thanks to the Internet of Things. Products like smart glasses introduce a variety of new capabilities. It is anticipated that the practice of placing new sensors on people's bodies will continue.

A system for building larger virtual networks and interfaces

Finally, a sophisticated infrastructure supports the Metaverse. This is frequently referred to as a nebulous system. However, most people are familiar with the elements of the Metaverse. For instance, everyone is familiar with wireless networks because they use their phones all the time. Not only will the quality of calls improve with 5G networks. Additionally, it offers faster data rates. As a result, many more people will be able to access the Metaverse's wealth of data.

A better interface is made possible by the improved functionality of mobile devices. However, it also makes it simpler to build VR and other display devices with small form dimensions. The Oculus Quest demonstrates that manufacturers are capable of developing VR systems that integrate mobile components. The extensive integration of features across several technologies aids in the infrastructure development of the Metaverse. Access to the Metaverse will become more prevalent as development continues.

Chapter 4

<u>Key technologies that power the Metaverse</u>

Technological evolution is crucial to enrich our experiences with extended reality and improve many aspects of the digital world. The Metaverse is a place of rapid technological development. Voice calls and text messages were the only forms of digital communication available to us just ten years ago; now, thanks to the Metaverse, we're part of the growing on an entirely different tangent.

Big technologies have combined to reproduce the Metaverse's fundamental idea and present a more developed 3D environment, which may provide users with an immersive online experience. Therefore, the public is being urged to prepare to take advantage of the Internet's upcoming evolution by Metaverse, which is now constructing a revolutionary change of the Internet.

Although the Metaverse is currently popular and more people are becoming aware of its uses, they still find it challenging to comprehend how it came to be. So, let's talk about all the technologies that are driving the Metaverse and enabling cutting-edge Metaverse projects. We will also quickly look at the Metaverse's history and development.

Why it is essential to understand Metaverse technologies

Businesses laud the numerous benefits of the Metaverse. However, to stay up to date with the most recent business trends and more effectively market their goods and services, they are prepared to invest in creating immersive environments that have been digitally upgraded.

Regardless of the nature of their industry, every firm should be aware of the numerous facets of the Metaverse because it is crucial to understand the Metaverse's role in business ecosystems, the effects it has on users, and how it links the real world and the virtual world.

Additionally, a technical grasp of the Metaverse's growth is necessary to explore the various facets of this open, shared, and persistent virtual reality for the continued creation of innovative and forward-thinking projects. Businesses can invite customers into the Metaverse and connect them to find more efficient growth opportunities by creating a project that is well curated. Companies must stay current with the Metaverse landscape to design and develop projects that significantly impact users and their target audiences.

Leading technologies that the Metaverse is powered by

Virtual reality (VR) and augmented reality (AR) innovations are frequently used to characterize the Metaverse. Metaverse development, however, goes beyond these two fundamental technologies. The best examples of using commonplace technologies to merge the physical and digital worlds are projects like Axie Infinity and Decentraland.

Blockchain is a critical component of the many technologies driving the Metaverse. Blockchain technology can make Metaverse projects consistent with Web 3, the upcoming Internet phase, by offering a decentralized infrastructure to Metaverse and encouraging the development of compelling use cases for its ecosystem. We will emphasize and investigate blockchain technology further in light of this.

Blockchain

1. Centralized Metaverses are possible, such as the ones used by Facebook and Microsoft. However, the centralized idea of the Metaverse is intended to be replaced with a decentralized and robust infrastructure. This is due to blockchains used in applications like NFTs and cryptos.

2. Blockchain is also essential to the growth of the Metaverse since decentralization is a key component of Web 3.0, being the next generation of the web. Therefore, even Metaverse initiatives must be developed as decentralized platforms to support a decentralized web. Projects in the Metaverse can benefit from decentralization thanks to blockchain technology. The decentralized components that blockchain provides for the Metaverse are as follows:

3. The decentralized blockchain database offers significantly more extended storage capacity than centralized storage. In addition, the

blockchain prevents space from running out for users in the Metaverse by leveraging unused hard disc space across the globe. Decentralized storage also aids in addressing centralized storage's other drawbacks, such as third-party involvement and potentially unethical behaviour like data theft and file tampering.

4. Exabytes of data are kept on the Metaverse. This causes concern about the security of data transfer, synchronization, and secure storage. Blockchain addresses these issues with its robust data processing, validation nodes, and decentralized storage capacity.

5. User data, virtual content ownership information, identification keys, and transaction information are all maintained on the blockchain in encrypted form. Through decentralized consensus, these data can be independently verified without the involvement of a third party. Only the nodes can access and validate these data to ensure trust and authenticity.

6. Numerous thousands of nodes can operate independently thanks to the decentralized Metaverse environment, allowing synchronization. A single source of failure, field programmable, or a delay in process execution cannot exist in the Metaverse.

7. The operation of the ecosystem's many regulations can be created and implemented by Metaverse via smart contracts. These agreements are self-executing, which implies that when a condition or set of conditions is satisfied, the agreement between the two parties is declared "Done."

8. Blockchain facilitates smooth interaction between different Metaverse initiatives, enabling them to carry out tasks, including sharing arbitrary data, associated information, and resources. A Metaverse project can be developed and run in various ways. However, the primary use cases that businesses should be aware of are listed below:

NFTs

A sizable portion of the total Metaverse initiatives in the works revolves around gaming. NFTs play a crucial role in these projects because they can provide immutable digital proof of ownership. In addition, NFTs control operations linked to digital asset trading, displaying users' unique avatars, creating a digital economy using cryptocurrencies, etc., when incorporated into the Metaverse.

Cryptos

Users of the Metaverse can engage in any activity that is feasible in the real world because it is a representation of it. Meeting new people, seeing new locations, and participating in trade activities are a few examples of these activities. All of these activities require a universal currency, which crypto can provide. In addition, crypto can be the driving force behind any financial exchange on a Metaverse. Users can use these funds to pay for services in both the virtual and physical worlds.

Identity verification for oneself

Any government-issued identification card, the social security number, passport, etc., is required for identity verification. Users' self-identity authentication functions similarly in the Metaverse. The distributed ledger, where nodes check these data, safely stores all user data, including body type, activity level, age, and other distinctive traits. It reduces the chance of fraudulent activity occurring inside the Metaverse. For instance, by taking on another person's identity, an avatar may engage in criminal activity.

The technology stack for developing blockchains

Front-end development is the branch of development that concentrates on the user-interactive graphic components of a decentralized application.

- React-native; JavaScript-based framework.

- HTML

- CSS

- Websocket

- Web3js

Backend development: Any blockchain-based solution's backend core is a smart contract, a piece of code governing a decentralized peer-to-peer application.

- Solidity, Rust, Node.js, React Native, DocumentDB, MongoDB, Auth0, and Web3js

1. Using VR and AR technology

The entry point to the Metaverse is augmented reality and virtual reality. These innovations are the enormous engines needed to power the Metaverse project, which can provide an immersive and more interesting three-dimensional virtual environment.

Users of virtual reality technology can connect with others in the Metaverse. One of the best instances of how VR technology is used to provide connected experiences is a shared meeting. With time, VR technology is integrating other significant technologies to advance its current use and improve Metaverse's capacity to deliver captivating and immersive 3D interactions.

However, augmented reality broadens the application of virtual reality. Consider a company that sells racing automobiles in a virtual reality showroom within the Metaverse. Only a virtual rendition of the cars in-store that is accessible in the Metaverse can be offered by brands through VR. However, augmented reality enables potential purchasers to test drive an automobile virtually before making a purchase.

Technology stack for the development of augmented and virtual reality

A feature-rich programming environment for real-time events, Unity supports 14+ platforms, encompassing desktop, smartphone, console, and VR devices, and multiple OS, including Windows, macOS, and Linux.

- Unreal Engine is an open-source, highly advanced real-time 3D editor created to produce lifelike and immersive experiences for various industries, including video games, media, and real estate.

- ARToolKit is an open-source, free library that enables the creation of real-time apps by providing features like automatic camera alignment, multilingual compatibility, and support for compasses.

- Google ARcore is a toolkit owned by Google that includes capabilities like motion tracking, light estimation, and environmental comprehension that work well with cutting-edge technologies like Unity and Unreal to facilitate the creation of AR and VR solutions.

- The Apple ARkit-Toolkit allows developers to change the 3D characters' facial expressions and image filters and improve lighting in digital items.

- With the help of the Maxst AR plugin, extended reality solutions can offer users an enhanced, lifelike experience by extending maps in accordance with camera movement and even beyond the camera.

1. Machine intelligence

Another significant factor that is really driving the growth of the Metaverse is artificial intelligence. Businesses can substitute activities with automated, computer-controlled ones by deploying AI. Fast computation, identity verification via facial recognition, statistics, and scaling better techniques are the most often used applications of AI for enterprises. Regarding the employment of AI in the Metaverse, it enables the creation of both 2D and 3D depending avatars based on the unique traits of users.

AI further enhances their realism by giving these avatars the ability to pick up on human body expression and conversational nuances. The same capability aids the Metaverse in developing automated assistance to answer users' questions through chat in the VR spaces.

Users of various languages can now explore the possibilities of the Metaverse and use its advantages thanks to the implementation of Ai to language processing on the Metaverse. Any language can be used to enter information, and AI will first break it down into machine-readable text before reading it and translating it into the target language.

Stack of technologies for developing AI

- **AWS sage maker:** The public cloud powered by Amazon web services offers crucial resources for developing, enhancing, and deploying AI and ML models.

- **TensorFlow:** A Google-developed open-source library for deep learning and conventional application development.

- **Scikit-Learn** is a crucial library designed to support Python language languages, which are frequently used in artificial intelligence and machine learning.

Reconstruction in 3D

The Metaverse should provide realistic virtual settings created using 3D reconstruction technologies because it represents the real world. By following the fundamental rules of physics and science, this technology enables the development of photorealistic things like buildings, a high-definition 3D environment, and actual locations inside the Metaverse.

Whereas 3D reconstruction in digital product development is a fairly well-developed idea, its method of application has changed over time. Real-world experiences are replicated virtually because to the work of businesses like real estate and e-commerce, which connect these areas into the Metaverse.

Through the use of virtual tours, businesses can assist prospective clients in viewing a property, evaluating it, and purchasing it for real estate. Real estate owners can sell their properties via Metaverse without meeting in person. Similar to this, online retailers have opened virtual shops where customers may browse products and decide wisely. For instance, a clothes company might provide showrooms and changing areas that imitate real shops.

Stack of technologies for 3D development.

In-Game Engine

An open-source, cutting-edge real-time 3D programming tool called Unreal Engine is used to build custom VR projects, interactive visualizations, and projects for the future generation of 3D technology.

The following features are available in Unreal Engine 5's most recent version:

- Open world toolset, next-generation real-time rendering, characters and complex animation tools built-in, an improved UI editor, and many other features.

Unity

With its tools, community, and support ecosystem, Unity is a complete real-time 3D, AR, and VR solution that helps experts create immersive 3D projects. You benefit from Unity by:

- The most sophisticated 3D editor is Unity Pro.

- Unity Gaming Services offers tools and resources to enhance your gaming projects.

- In-house artist tools and well-curated documentation are also available.

CRYENGINE

Famous 3D development tool CRYENGINE was created to help game makers. This tool also functions as a potent VR development tool with the following capabilities and compatibility for numerous platforms:

- Creation of gorgeous graphics and believable personalities.

- Encourage the creation of user-friendly, feature-rich VR apps.

- Provides integrated audio solutions that allow for the development of incredibly immersive digital experiences.

Internet of Things (IoT)

IoT is a technology that employs sensors and other devices to act as a conduit between the physical world and the Internet. IoT strengthens the link between both the Metaverse and physical things or gadgets in the physical world. These gadgets can effortlessly transmit and receive information once linked to the Metaverse, improving how accurately the physical world is replicated.

IoT also gives the Metaverse access to real-time information gathered from the outside world. By making use of such information, the Metaverse can improve the integrity of the events taking place in its virtual world by rendering them pertinent to the circumstances or environment in the actual world.

We might refer to these IoT jobs in the Metaverse as complements to one another. Considering how they keep developing new ways to power each other, some analysts have even dubbed them a technology twin. Both contribute to the world's ability to make better data-driven judgments more quickly and with fewer resources.

The production of real-time simulation in the Metaverse is made possible by the use of IoT, which may symbiotically connect the 3D environment to numerous real-world devices. Furthermore, IoT can combine with AI and artificial intelligence technologies and efficiently manage the data for future Metaverse optimization.

IT stack for Internet of Things development.

MQTT

A common messaging protocol is made to connect remote devices with a small network bandwidth and code.

IoT hub for Azure

Azure IoT hub, also referred to as Microsoft's IoT adapter for the cloud is a wholly-owned cloud service provided by Microsoft that enables a smooth connection between several IoT devices and a backend solution.

AMQP

A sophisticated standard protocol that provides synchronization between a variety of systems and apps, allowing them to cooperate and guarantee that the information gets to the intended customers.

Lorawan

A provided support with low-power capabilities called Lorawan makes IoT deployments more effective and guarantees continuous connection between an end node and integrated service devices.

The development of the Metaverse has not yet reached its full potential. However, tech professionals are investigating methods and evaluating various technologies to create concepts that will give the world more exciting and workable Metaverse initiatives.

Axie Infinity, Sandbox, and Decentraland are a few well-known examples of modern live Metaverse initiatives. These products' creators made excellent use of notable Metaverse technologies which include, but are not limited to blockchain, NFTS and cryptography.

According to leading tech bods, the Metaverse will inevitably be empowered by several new technologies in the future, supporting the creation of a variety of practical use cases that will increase the Metaverse's capacity to provide real-life functionality. In summary, the Metaverse is predicted to present businesses and regular consumers with enormous prospects. So, it will influence how the digital world develops.

Chapter 5

The Metaverse and Crypto

As you have learned, all you need to access the Metaverse is a personal computer, iPad, or smartphone with a reliable Internet connection. However, the right cryptocurrency must also be present in each person's virtual wallet; therefore, this is insufficient.

What is it?

The virtual wallet functions similarly to the physical one we keep in our pockets for holding cash. The main distinction is that since money is fictitious in the world of Metaverse, it is referred to as "cryptocurrency." However, you can engage in all the typical real-life actions with these currencies, including buying, selling, storing, and keeping anything to enhance or reduce their worth. There is no distinction between the economic transactions we can carry out using euros, dollars, or virtual currencies, to put it simply.

A Virtual Currency

Cryptocurrencies are digital money that may be purchased and traded in the Metaverse. As a result, they are payment methods that make trading entirely secure, quick, easy, and private. Therefore, exchanges can take place anywhere in the world. This enables you to go above any restrictions and geographical limitations in this case. The Bitcoin system does not contain financial intermediaries that eventually lead to systemic imbalances. Instead, this occurs in traditional markets, when governments consistently exercise some form of social control. In the end, the only parties engaged are the brokers who cope with the acquiring and selling of virtual currencies, naturally in exchange for payment of a certain amount of money.

You shouldn't, however, assume that simply though these coins are not actually real, there are a limitless number of them and that they are constantly being sold on the market. Contrarily, a maximum quantity has been predetermined for each virtual currency, beyond which you cannot go. This prevents the development of extreme inflation, and cryptocurrencies may experience a drop in value (at least for most of them).

You can compare this idea with the actual coins and banknotes we use daily to grasp it better. Because of inflation, the Bank of Italy cannot constantly issue 5euro banknotes without risking our country's financial collapse. Those bills would ultimately amount to nothing more than worthless pieces of paper.

Although Bitcoin was the first cryptocurrency available to the general public, many different kinds of cryptocurrencies exist. Depending on their formulation or code design, application or use case, and other criteria, we can distinguish at least four different sorts of cryptocurrencies.

Coins, alternative coins for payments, security tokens, non-fungible tokens (NFTs), utility tokens, and other types of tokens are possible.

This tutorial explains the many kinds of cryptocurrencies and tokens. We also include details on the many sorts of cryptocurrencies, how they are used, along with examples.

What Makes Cryptocurrencies Different?

Despite being used to describe all types or varieties of digital or cryptocurrency, the term "cryptocurrency" is frequently used interchangeably with "coins." This is because they are commonly considered as such, even though most of them—aside from Bitcoin— do not function as a unit of account, a store of value, or a medium of exchange.

Coins can be distinguished from altcoins, though. In addition to Bitcoin, all other cryptocurrencies viewed as alternatives to Bitcoin are referred to as altcoins.

Coins: Since coins are built on their own blockchain, they can be distinguished from altcoins. They serve as both the native token and the fuel or gas payment token on such a blockchain, albeit a blockchain might have the gas paid in a different cryptocurrency. Bitcoin which is run on the Bitcoin blockchain and Ether, or ETH, on the Ethereum blockchain are two really good examples. Building or creating a cryptocurrency begins with or follows the creation of a blockchain.

Altcoins: While these can technically be considered coins, they are all considered alternatives to Bitcoin, the original cryptocurrency. Apart from Ethereum, most of the original ones also referred to as "shitcoins," were forked from Bitcoin. These include Auroracoin,

Litecoin, Dogecoin, Peercoin, Namecoin, and Peercoin. Nevertheless, other cryptocurrencies have their own blockchains, including Ethereum, Ripple, Omni, and NEO. Some do not.

Tokens: In a blockchain, a token is a digital representation of a particular asset or utility. All the tokens can be referred to as altcoins; however, they differ from one another by existing on top of a different blockchain and not being a part of that network natively.

We can move some of them from one chain to another because they are programmed to support smart contracts on blockchain networks like Ethereum. The tokens can run independently of a third-party platform since they are integrated into self-executing computer programs or algorithms. They can also be traded and are fungible. They can serve as a representation for commodities, rewards points, and even other cryptocurrencies.

The developer must adhere to a predetermined template while creating or coding a token. The developer does not need to rewrite the blockchain in code completely. All they have to do is adhere to a pre-established standard template. As a result, the process of creating a token is quicker.

Initial Coin Offerings, also known as ICOs and IEOs, were once the main ways to distribute and raise money for companies creating tokens. They can, however, be released without IEO or ICOs.

Various Forms of Cryptocurrency

Utility Tokens

Utility tokens are quite simply digital units which reflect a value on the blockchain. They are sometimes regarded as vouchers or coupons. To explain it more clearly, the token grants specific access to a good(s) or service(s) offered by the token issuer. By purchasing that token, a person can acquire access to the product or service and exchange it for a specified access value.

- The holder does not acquire ownership but rather the right to a good or service of an equivalent value. For instance, as long as they own the tokens, they can access the good or service for a reduced cost or for free.

- In some jurisdictions, a cryptocurrency classified as a utility token is exempt from financial regulation.

- The main concept is that they are not financial items, and that the possessor may lose all of their money if they entirely lose value.

Because they are not expected to be regulated, utility tokens are easier to understand from a regulatory standpoint. Applications include access to decentralized storage in a decentralized storage network, incentives tokens, and as money for a blockchain. The token holder is not holding an equivalent of a stock, bond, or other asset regulated under financial acts.

Funfair, Basic Attention Token, Brickblock, Timicoin, Sirin Labs Token, and Golem are a few examples of utility tokens.

Security tokens

These cryptocurrencies are securitized and derive their value from an exterior asset that can be sold as a security in accordance with the financial regulations in force at the time. As a result, they are used to securitize the tokenization of bonds, stocks, real estate, and other international currencies from the known real world.

- Due to the nature of these transactions, financial authorities must exercise control over and regulate their exchange, issuance, deals, value, steganography, backing, and trading to safeguard user investments.

- In this situation, the regulation is in place to protect user cash and resources and to hold entrepreneurs accountable.

A stake, share of stock or equities, voting rights, and a right to the income in the asset represented are all represented by security tokens. Owners or holders share in the profits made by issuers or managers who make decisions and take action. They can be used in situations when investors require instantaneous settlement, openness in management, disreputability of assets, etc. They are distributed through security token offerings (STOs).

Security tokens are additionally separated into:

• **Equity tokens:** Similar in structure and operation to traditional equities, but with digital ownership and transfer. Dividends from administrative and issuer choices and actions are due to investors. Debt tokens represent short-term loans with fixed interest rates.

• **Asset-backed tokens:** These have tangible assets like commodities, works of art, real estate, or even carbon credits as their underlying worth. They possess qualities similar to gold, metal, oil, etc. They can be exchanged, etc.

Sia Funds, Bcap (Blockchain Finance), and Science Blockchain are a few examples of security tokens.

Money Tokens

Payment tokens are used to transact directly between a buyer and seller of products and services on the various digital marketplaces, as opposed to going via an intermediary, as we do in traditional banking and finance. The bulk of cryptocurrencies and tokens fit within this category, whether they are security or utility tokens. Nevertheless, not all utility tokens can also function as payment tokens.

- Payment tokens are not securities and cannot be purchased as such. As a result, they are not considered asset securities subject to financial regulation.

- They might or might not ensure holders' access to any present or future goods or services.

Monero, Ethereum, and Bitcoin are examples of money tokens.

Exchange Tokens

Exchange tokens, which are given the term because they are issued by and used on the crypto exchange, which is an online market for buying, selling, and exchanging tokens, may be disputed.

Even though they can be utilized outside their original exchange contexts, we mainly employed them to facilitate token trades or pay for gas on these exchanges.

- They may be issued by centralized exchanges that have their own blockchains or decentralized platforms.

- They can be used to pay gas or other costs more cheaply, increase liquidity, offer free discounts, administer blockchains by granting voting rights or grant access to specific crypto exchange services.

- Exchanges utilize them to entice people to participate in the projects to increase liquidity.

Binance Coin, also known as the BNB token, Gemini USD, FTX Coin for the FTX Exchange, OKB for the Okex Exchange, KuCoin Coin, Uni Token, HT for the Huobi Exchange, Shushi, and CRO for Crypto.com are some examples of exchange tokens.

Un-transferable Tokens

A non-fungible token is a blockchain-based digital proof of ownership for a singular, non-replaceable item or one that cannot be traded with another asset.

It is created using the same technology as other types of tokens. Still, it is primarily used to represent a work of art, images, audio, video, collectables, real estate, virtual worlds, memes, GIFs, online downloads like posts and tweets, fashion, music, paintings, drawings, pornography, academic content, political items, films, memes, sports, games, or digital files of value that are stored on the blockchain.

- The Ethereum blockchain saw the creation of the first NFT in 2015.

- The digital signature was made in a way that prevents it from being modified for another.

- They permit the owner to possess a unique object with a restricted supply, original design, or edition.

- The issues can be limited edition or impossible to replicate or reproduce due to their high value. The best NFTs are ones in which only a small number of people can own an original.

They can be purchased and sold in NFT markets like OpenSea, Rarible, Foundation, and Decentraland. It mainly assists artists, makers, and collectors in selling their goods.

• The applications include fame, monetizing goods, for royalty payment so that artists receive a percentage of the sale whenever the art is sold to a new user, partial ownership of land and costly assets, auctioneering to raise money like Charmin and Taco Bell auctioneering of themed NFTs, creating special moment memories or preserving historical events, for market motives like trading, and celebrity issuing.

Logan Paul's videos, Jack Dorsey's early tweets, EVERYDAYS: the First 5000 Days drawings by Mike Winklemann, best known as "Beeple," and a few crypto kittens are examples of NFTs.

Tokens

Decentralized finance relates to money applications, or dApps, built on the blockchain or distributed record, which makes them scattered. Those that render financial and money power directly to the user while allowing them to transact globally with peers-to-peers methods and access to global markets. Anyone with Internet access can access these DeFi apps. Each DeFi app runs on a token economy supported by a native token. These tokens are programmable currency where developers may add logic to how transactions and payments are made.

The Ethereum blockchain is presently the foundation for the majority of DeFi currencies. Other blockchains that allow DeFi include Cardano, IOTA, Tron, Polygon, and Stellar.

- Through these tokens, users can send and receive money, lend money, borrow money, earn interest, save money, grow and manage their portfolios, invest in securities, stocks, and funds, send and receive money, trade money on decentralized exchanges, buy and sell assets, invest in assets, and more.

Solana, Chainlink, Uniswap, Polkadot, Aave, and other well-known decentralized finance tokens are examples. Decentralized lending apps, decentralized exchanges, decentralized storage sharing, etc., are some examples of DeFi application categories. Smart contracts, *which enable anybody to establish, write, program, and execute transaction rules based on specific circumstances and have actions performed when those requirements are met, are the most potent component of DeFi tokens.*

Fiat and other stablecoin types

These tokens are of a stable value, as their name implies, meaning that their value is reasonably predictable in that it essentially never changes. In addition, stable tokens, or stablecoins as they are more commonly known, are backed by an asset with a relatively consistent value, such as cash. So, we have stablecoins pegged to the dollar and the euro and tokens backed by commodities like oil, gold, and other precious metals.

Stable tokens are backed on a specific ratio, and the asset supporting them must be retained in reserves in accordance with the defined ratio. They enable the world to eliminate volatility in assets or other virtual currencies. Some stablecoins are backed by fiat, cryptocurrency, commodities, and algorithmic stablecoins, which rely on rules and software to keep their value pegged to fiat or another asset.

Stablecoins include Paxos and Tether, which are all backed by USD cash in a 1:1 ratio, along with TruSD, Gemi Dollar, and USD Coin as stablecoins backed by gold, Kitco Gold, Tether Gold (XAUT),

DigixGlobal (DGX), and Gold Coin (GLC) are also available. In addition, there are other stablecoins supported by algorithms, such as Ampleforth (AMPL), DefiDollar (USDC), Empty Set Dollar (ESD), and Frax (FRAX).

Tokens backed by assets

The term "asset-backed tokens" refers to a class of cryptocurrencies where the underlying value is supported by a real-world asset, such as cash, stocks, bonds, property, gold, and precious metals. On blockchains, they are utilized to represent and trade the value of these underlying assets digitally.

Due to the way transactions involving the underlying assets are conducted, most of these are provided as security tokens. Therefore, most of them are distributed via the Equity Tokens Offer (ETO).

Depending on the issuer, they could be guaranteed at any ratio.

- Tokens backed by precious metals are the gold-backed PAXG and DGX. Visit our other lesson for more information on alternative gold-backed tokens.

- Tokens backed by company shares enable the tokenization of firm shares and their trading on cryptocurrency exchanges. Examples include Slice, The Elephant Private Stock Coin, Quadrant Token (which tokenizes the Quadrant Biosciences Inc. equity), BFToken, The Dao, and RRT Token.

- In tokenized products, the value of items is represented by tokens, which also go by the name of crypto-commodities. These commodities include wheat, sugar, natural gas, oil, and renewable energy sources.

Examples of tokens backed by assets are OilCoin, which tokenizes reserve oil barrels, Petroleum Coin, Ziyen Inc. Oil token, etc. Energy was tokenized via the Energy Web Token (EWT), Green Energy Token by WPP, etc. Tokenization of wheat using the Wheat Token Coin

Privacy tokens

These cryptocurrencies are used for privacy applications, as their name implies, because they have superior privacy-promoting programming to Bitcoin and other popular cryptocurrencies. Better anonymity in crypto transactions is necessary for various reasons, including the right to privacy, security investigations, and extremely sensitive transactions, yet they are also utilized in fraud and crime.

- These cryptocurrencies use a variety of strategies, like currency mixing, offline transactions, and anonymity mechanisms like CoinJoin, to ensure transaction privacy. This is in addition to methods used in mainstream cryptos, such as blockchain encryption and a lack of attaching real names to crypto addresses.

The privacy tokens Monero, Zcash, Dash, Horizen, Beam, and Verge are a few examples.

Bitcoin vs. Ethereum

Overview

The second-most well-known digital token after Bitcoin is Ether (ETH), the money of the Ethereum network (BTC). Therefore, the foremost cryptocurrency by market capitalization (market cap), Ether, is a natural subject for comparison with Bitcoin.

Bitcoin and Ether are comparable in many ways: Each is a digital currency that can be exchanged through online marketplaces and is kept in different kinds of cryptocurrency wallets. Both coins are decentralized, meaning no central bank or other institution issues or controls them. Both make use of blockchain, a distributed ledger technology.

But there are also sever significant differences between the two most widely used cryptocurrencies by market cap. We'll examine the distinctions and similarities between Bitcoin and Ether in more detail below.

Basics of Bitcoin

Launched in January 2009, it introduced a revolutionary concept laid out in a white paper by the enigmatic Satoshi Nakamoto: Bitcoin promises to be an online currency safeguarded without any central power, in contrast to currencies produced by governments. There are just balances linked to a public ledger that has been cryptographically safeguarded; there are no actual Bitcoins.

Despite not being the first attempt at online money of this kind, Bitcoin was the most profitable in its early efforts and has come to be considered a forerunner of sorts to almost all cryptocurrencies that have been produced over the previous decade.

The idea of virtual, decentralized money has become more and more popular among authorities over time. Cryptocurrency has clearly carved out a niche for itself and continues to coexist with the financial system despite being frequently questioned and contested. This is even though it isn't a formally recognized mode of payment or store of value.

Basics of Ethereum

Beyond only enabling digital money, blockchain technology is being leveraged to build apps. For example, the largest and most established open-ended decentralized software platform, Ethereum, was introduced in July 2015.

With the help of Ethereum, decentralized apps (dApps) and smart contracts may be developed and operated without interruption, fraud, centralization, or outside influence. Developers can create and manage distributed apps using Ethereum's built-in, blockchain-based programming language.

Wide-ranging potential uses for Ethereum are enabled by its native cryptographic coinage, Ether (commonly abbreviated as ETH). Ethereum started an ether presale in 2014 and was incredibly well welcomed. 8 Developers utilize Ether to create and execute apps on the Ethereum platform, acting as the network's equivalent of fuel for processing commands.

The two primary uses of Ether are to run apps on the Ethereum network and to trade it as a cryptocurrency on exchanges in the same way other cryptocurrencies are exchanged. In addition, people "all over the world use ETH to make payments, as a store of wealth, or as collateral," claims Ethereum.

Significant Differing

Although the distributed ledger and cryptography principles underlie both the Bitcoin and Ethereum networks, there are significant technological differences between the two. For instance, data attached to transactions on the Ethereum platform may contain source codes, whereas data attached to transactions on the Bitcoin network is often merely for note-taking. The block time (an ether transaction is validated in seconds, in contrast to minutes for Bitcoin) and algorithms used by each network—SHA-256 for Bitcoin and Ethash for Ethereum—are other differences.

Proof of work (PoW), a consensus protocol that both Bitcoin and Ethereum now use, enables the nodes of the respective networks to concur on the status of all data stored on their blockchains and guards against some sort of economic attack on the networks.is, or by now has switched to (in 2022) as part of its Eth2 upgrade, a series of connected updates that will increase Ethereum's scalability, security, and sustainability. Proof of work requires a heck of a lot of computer power and is hence very energy-intensive is one of its main detractors.

Proof of stake replaces mining with staking, which uses less energy, and validators instead of miners who stake their Bitcoin holdings to enable the creation of new blocks.

However, the primary objectives of the Bitcoin and Ethereum networks differ, which is more significant. While Ethereum was designed as a platform to enable immutable, automated contracts and applications via its own currency, Bitcoin was founded as an alternative to the monetary system and strived to be of a means of exchange and a store of value. Both Bitcoin and Ether are digital currencies, but Ether's primary goal isn't to establish itself as a competing currency system. Instead, it wants to make it easier and more profitable for the Ethereum smart contract and decentralized application platform to function.

Another application for a blockchain that enables the Bitcoin network and, in theory, shouldn't actually compete with Bitcoin is Ethereum. However, due to its rising popularity, ether is now in direct conflict with all other cryptocurrencies, particularly in the eyes of traders. Ether has consistently placed second to Bitcoin in rankings of the major cryptocurrencies by market cap since its introduction in the middle of 2015.

What is the primary application difference among Bitcoin and Ethereum?

As a replacement for conventional currencies, Bitcoin is primarily intended to serve as a medium of exchange for goods and services of wealth. Ethereum is a programmable blockchain with use cases in DeFi, smart contracts, and NFTs, among others.

Chapter 6

Understanding Blockchain

You should carefully consider a few key factors to make this decision wisely and intentionally and prevent regret in relation to purchasing cryptocurrency for use in the Metaverse. First, you must think about where you will be able to purchase the money and the accompanying fees. You must know the minimum amount to buy, the commissions' details, whether they are fixed or variable, the required initial investment, and the platform to utilize.

Second, I advise you to consider the possible uses for that specific currency because you risk purchasing one that doesn't apply to your Metaverse. You can conduct economic and commercial transactions using payments made using cryptocurrencies on the Internet and, consequently, within the Metaverse.

A "block" is referred to whenever one of these deals is approved and confirmed. When it receives an absolute majority from every node in the network, validation takes place. As soon as this occurs, the block is placed on the so-called "blockchain" (a chain made up of all the blocks that have been verified).

What function do blockchain and cryptocurrency serve in the Metaverse?

The establishment of a digital economy is aided by the availability of Bitcoin in the Metaverse as opposed to the web's PayPal and credit card payment options. Blockchain has shown to be helpful for six key Metaverse categories: digital collectability, recognizing, governance, transparency, and interoperability. In addition, blockchain technology offers a transparent and affordable alternative, making it suitable for the Metaverse.

What does the blockchain-Metaverse relationship's future hold?

There is still much to be developed in the Metaverse. You'll observe that the games' mechanics, visual style, and sensory experience are all relatively simple if you play them for yourself. However, many are not yet ready for testing and are still in the planning stages.

One thing is sure, though: increasingly, more new ventures are being launched. Development is advancing quickly, regardless of whether it's a little Metaverse crypto initiative or a major gaming firm. The projects mentioned above are only the beginning; keep up with the Metaverse by constantly checking for upgrades and news.

Why a blockchain is essential for the Metaverse to function

Blockchain is a brilliant system that permanently logs transactions, often in a ledger, which is a decentralised and open database. The most popular blockchain-based crypto is of course Bitcoin. Every time you purchase some Bitcoin, that transaction is added to the blockchain which distributes the record to thousands of different computers worldwide.

It is incredibly difficult to trick or manipulate this decentralised recording system. Furthermore, as opposed to previous banking records, blockchains like Bitcoin and Ethereum are totally transparent in that all transactions that are carried out are visible to anybody online.

As with Bitcoin, Ethereum uses a blockchain, but Ethereum can also be programmed using smart contracts, which are effectively blockchain-based software routines that execute when certain conditions are satisfied. For instance, you could personally use a smart contract on the blockchain to prove you are the rightful owner of a digital asset (such as a work of art or a piece of music) and prevent others from claiming ownership of it on the blockchain — even if they save a duplicate to their computer. Crypto assets are things that can be possessed in a digital format such as money, stocks, and art.

Nonfungible tokens on a blockchain are things like music and art (NFTs). However, in contrast to fungible goods like money which are interchangeable and each worth the same as others, nonfungible goods are singular and cannot be replaced in any shape or form.

Most significantly, you may utilise a smart contract that specifies that you are prepared to sell your work of digital art for US $1m in ether, the cryptocurrency used by the Ethereum network. The artwork and the ether instantaneously transfer ownership to one another on the blockchain when either of us click "agree." No bank or third-party escrow is required, and if either of us were to contest this transaction — say, if you insisted that I only paid $999,000 — the other might just as easily point to the distributed ledger's public record for verification.

So, what exactly is the connection between the Metaverse and this blockchain crypto-asset stuff? Absolutely everything! Thanks to the wonderful blockchain, you can first possess some digital commodities in a virtual environment. That NFT will be yours - not only in the physical world but also in cyberspace.

A single organisation or group isn't creating the Metaverse. Different organisations will create various and numerous virtual worlds. In future, these worlds will be able to communicate easily with each other, establishing the Metaverse. People will want to take their belongings and possessions with them as they switch between their virtual settings, such as from Microsoft's world to Decentraland's. The blockchain will be able to confirm ownership proof of your digital products in both imaginary spaces if two of them are interoperable. In a nutshell, you will easily be able to access your crypto assets as long as you can access your virtual world wallet.

Keep your wallet close by

In the Metaverse, you will clearly want to carry cryptocurrency. Therefore, your digital items that are exclusive to the Metaverse - your avatars, animations, clothing, virtual accents, and weaponry, will be safely stored in your cryptocurrency wallet.

How will people use their cryptocurrency wallets? For shopping, among other things. You will be able to obtain conventional digital items such as music, games, apps and movies just like you already can do on the Internet. The ability to examine and "hold" 3D models of the products that you are shopping for will enable you to make more calculated or informed decisions. You will also be able to buy items from the actual world in the Metaverse.

Additionally, cryptos wallets can link to real-world identities, which might assist in expediting transactions that need full legal proof, such as buying a real-world vehicle or property. It is similar to how you can use your old leather purse or wallet to carry your ID in. You won't need to remember tedious login details for every website and a virtual world you visit, because your identification will be linked to your wallet. All you will need to do is connect your wallet with a single click to log in. Access limitations to age-restricted parts of the Metaverse will be made possible through your ID-associated wallet.

Large enterprises

Finally, businesses will undoubtedly want to participate in the Metaverse if it proves profitable. Although the decentralised structure of blockchain may lessen the need for intermediaries in financial transactions, businesses will still have numerous chances to make money—possibly even more than in economies as they exist today. In addition, large platforms will be provided by companies like Meta for individuals to work, play, and gather.

Major companies like Adidas, Dolce & Gabbana, Coca-Cola, and Nike are joining the NFT fray. In the future, you might also be able to acquire ownership of a linked NFT in the Metaverse when you purchase a real-world item from a business.

These are just some examples of how Metaverse business models and the real world may be able to happily coexist. Such instances will inevitably become more complicated as augmented reality technology becomes more prevalent, further fusing together the Metaverse with the physical world. The Metaverse itself may not be here just yet. Still, technology underpinnings like blockchain and crypto assets are steadily being built, laying the groundwork for a seemingly pervasive virtual future that is soon to arrive in a "verse near you."

Chapter 7

Metaverse and NFT

A brand-new class of digital assets built on blockchain technology is called non-fungible tokens, or NFTs. NFTs essentially serve as a network's representation of assets. For instance, you may produce original digital art and turn it into an NFT. The piece of art would continue to exist on a Bitcoin blockchain with access restrictions set by the owner.

On a blockchain network, real estate can similarly be represented by NFTs. In the Metaverse vs NFT contrast, what makes NFTs so unique? The distinctive feature of NFTs is that they reflect distinct ownership titles to a particular physical or virtual item. If you are the sole rightful owner of an NFT, no other person may use their ownership rights in connection with the asset.

The most frequent distinction between NFT and the Metaverse that people make centers on the blockchain, which is present in both Blockchain technologies and is used for more than only cryptocurrency transactions as a decentralized, peer-to-peer network. It can function as a record of transactions involving a particular asset and aid in tracing the path taken by the NFT through various transactions. Who purchased the NFT? Are the NFT going to be sold again in the future? As the NFT's owner, you may readily find whatever information you need regarding the NFT.

NFTs are distinctive assets since you cannot exchange them. Non-fungible tokens can only be traded for other non-fungible tokens, which provide the security of a single proof of ownership. For example, a few sets of brand-new trading cards cannot be exchanged for a vintage trading card. The term "non-fungible" quite simply refers to the fact that each NFT has unique characteristics that set it apart from the others.

Actual Applications of NFTs

The comparison of NFTs and Metaverses would also highlight the various examples of non-fungible tokens found in musical compositions and visual works of art. It could be helpful to get a clearer sense of how the non-fictional worlds (NFTs) vary from the Metaverse if one looked at some examples of the NFTs. You might be familiar with some well-known examples of NFTs, such as the CryptoPunks and the BAYC NFTs.

Because they are quickly being recognized as the most engaging profile images available on the Internet, Bored Apes have become extremely popular with famous people. Similarly, the usage of NFTs in other scenarios implies a representation of ownership over property resources. The various characteristics of these technologies support the varied applications of NFTs. When you purchase an NFT, you will be given the ownership certificate corresponding to the underlying asset.

In comparing NFTs and the Metaverse, you can locate the next pointer by looking at the accessibility of NFTs. Where precisely can you find them? The ease with which you can access non-financial resources is unquestionably one of the essential factors determining whether you will have those resources available to you when you require them. One of the most important benefits of having access to NFTs whenever you need them is the vast number of available NFT marketplaces. Some of the most well-known NFT platforms, such as OpenSea, give users access to all different kinds of NFTs along with information about those NFTs in great depth. You can make purchases of NFTs using cryptocurrencies precisely the same way you would make purchases of other crypto assets.

NFTs and the Metaverse differ from one another

The two trendiest terms in the developing web 3.0 landscape at the moment are NFTs and the Metaverse. The question "Are NFT's and Metaverse the same?" is one that nearly everyone in the tech community wants to know about due to the tremendous and unprecedented rise in both technologies' popularity. Instead, it's critical to consider the distinctions between the two terms. By examining their differences, you may better grasp how NFT and the Metaverse fit into the more extensive web 3.0 ecosystems.

I imagine that many of you thought that the Internet as we know it now was the definitive version. However, web 2.0 (or the Internet as we know it today) is not without its challenges. For instance, centralized businesses control user data, a significant setback for consumer privacy.

The ideas of Metaverse and NFTs are revolutionizing the destiny of the Internet. The critical distinction between NFT and Metaverse is between the two concepts' fundamental definitions. The Metaverse is a separate virtual environment, whereas non-fungible tokens are a subset of virtual tokens. Here is a thorough comparison of their differences.

Foundations

The definitions of NFTs and the Metaverse offer sufficient justification for a helpful comparison. However, you should be aware that blockchain technology is the fundamental pillar of the comparison between NFT and the Metaverse. Because it is essential for creating smart contracts, which regulate ownership and transactions with NFTs, blockchain is an integral component of NFTs.

On the other hand, the Metaverse is a sizable universe built around the idea of developing an open, shared, persistent, and intensely participatory Internet. Immutability, non-fungibility, and security are characteristics of non-fungible tokens. In addition, the Metaverse offers various features, such as decentralization, user identification, creative economy, and experiences.

Origins

The invention of CryptoPunks in 2017 marked the beginning of NFTs, which have been around for a while. The CryptoKitties collection also made headlines around this time about Ethereum network congestion. The history of NFTs points to the potential for developing new blockchain-based assets that signify exclusive ownership. As a result, the introduction of NFTs may have spurred fresh ideas for promoting asset ownership decentralization.

Comparing the origins of the Metaverse and NFT reveals the ideas and objectives driving each technology. The Metaverse does not have a single aim, speaking of which. Decentralization is enhanced, and numerous additional use cases are made possible. The Metaverse is rooted in a science-fiction book that presents it as a haven from the real world.

Usability

The following obvious distinction between non-fungible tokens and the Metaverse would be their usefulness. How straightforward is it to use NFTs or Metaverse platforms? There are several venues available for finding the best NFTs, like NFT markets. For example, OpenSea is now the largest NFT marketplace where you may browse through all of the NFT's specifications before purchasing them.

Additionally, you would discover that numerous systems make it simple to enter the Metaverse. Users can access the Facebook Meta platform, the Sandbox Metaverse, the Roblox game Metaverse, and other Metaverse platforms. In addition, you can access any Metaverse platform with just a set of your preferred VR or XR gear.

NFTs' impact on the Metaverse

You may have discovered the various ways that NFTs can contribute to the creation of the Metaverse in your search for the answers to the question, "Is NFT part of the Metaverse?" Nevertheless, it is crucial to recognise how NFTs have altered the Metaverse's basic structure. Furthermore, you must be aware that NFTs may cause disruptions in the user engagement, transaction, and socialising patterns typical of social networks in the Metaverse. How, therefore, might these impacts be broadly translated onto the Metaverse? Here are some critical aspects of the future Metaverse NFT interaction.

The Path to a Just and Open Economy

Individual users and businesses can easily represent their tangible assets and solutions in a decentralised digital ecosystem. Furthermore, by using cutting-edge gaming technologies in tandem with interoperable blockchain games, the Metaverse may become more accessible to a broader range of real-world assets.

New models, such as the play-to-earn gaming paradigm, would increase the prominence of NFT in the Metaverse. It provides players with opportunities to be empowered and the option to use NFTs to increase involvement in the Metaverse. Additionally, play-to-earn games offer a fair gaming experience by giving players total ownership and control over all assets.

It was nigh on impossible to miss the importance of play-to-earn gaming guilds in fostering the development of the NFT Metaverse interaction. The guilds act as middlemen, buying in-game NFT resources like assets and lands. Then, they lend the property and assets to gamers so they can use them in other virtual worlds to generate income. The play-to-earn guilds would only retain a negligible portion of the profits in exchange. As a result, by utilising NFTs, you can find the ideal foundation for a just and open economy in the Metaverse.

By providing a head start to players without an initial investment, guilds could lessen the entry barrier for games that need play to earn money. As a result, you can see that the Metaverse has the potential for a just economy in which everyone can participate. On NFT marketplaces, users could freely exchange their NFT assets, such as in-game collectables and virtual property.

Experiences of a New Generation in Terms of Community, Society, and Identity

The effect of NFT Metaverse initiatives would also play a big part in modifying the identities, social interactions, and community-building opportunities available to users in the Metaverse. For example, users with NFT assets can demonstrate their support for a particular project or express their ideas concerning the virtual and real worlds. As a consequence, NFT owners with similar interests could come together to form communities to exchange information and work together on the production of content.

The example of NFT avatars recently gained popularity demonstrates how the relationship between the Metaverse and the NFT affects the actual world. A player's authentic self and the self they envision themselves to be are represented by their NFT avatar. The NFT avatars of the players might be used in the role of access tokens, allowing them to move freely between the various parts of the Metaverse. You might think of non-fungible tokens (NFTs) as an extension of the users' real-life identities, with full ownership, control, and flexibility over the process of constructing their virtual identities.

Users could earn virtual membership to various experiences, both in the real world and in the Metaverse, using NFT avatars. As a result, the integration of NFTs with the Metaverse has the potential to enhance the social and communal aspects of the user experience. The potential of NFT avatars is further demonstrated by their applications in the Metaverse, which include the founding of start-ups and the production of content.

The fact that blockchain enables transparency and immutability makes the role of NFT in the Metaverse abundantly clear that these characteristics significantly impact the fair and open economy in the Metaverse. Now, NFT scarcity and their on-chain value would be determined by the fundamental law of supply and demand. Therefore, you could not identify any opportunities for artificial value inflation. Thus, it is clear how the Metaverse and NFTs combine to establish a trustworthy and equitable economy.

Emerging Patterns in the Real Estate Market

Virtual worlds result in a significant amount of virtual space and real estate. Therefore, you could fully own virtual areas in the Metaverse by using non-fungible tokens (NFTs). Furthermore, users would have an easier time establishing ownership of the item with the assistance of the blockchain, which would also allow for the development of the virtual real estate.

One of the most famous applications for such NFT Metaverse initiatives is the purchase and resale of virtual land for financial gain. In addition to creating various structures, such as online shops or event hosting, you may also generate passive revenue by renting out land.

The most well-known illustration of how the digital real estate market works may be found in Decentraland, which exists in the Metaverse. Not too long ago, Decentraland hosted a virtual fashion exhibition, and they did it in conjunction with Adidas. During the event, the auction of fashion designs was presented as NFT. Therefore, there is no need to be reluctant when considering the prospects of holding auctions for virtual locations in the Metaverse. The increasing popularity of virtual real estate has also piqued the curiosity of musicians who want to maintain ownership over their work. In the future, the term "digital real estate" will be used to refer more generally to the ownership of digital assets, and each NFT holder will have their own place in the Metaverse.

How to Earn Money with Unique Assets

Investing in non-fungible tokens (NFTs) and real estate in the Metaverse

Landowners in name only

As was previously indicated, one way that individuals make use of NFTs in the Metaverse is to purchase virtual land, such as LAND, a piece of digital real estate located in The Sandbox. As an alternative to a physical deed, these virtual spaces use non-fungible tokens, or NFTs, to reflect ownership of specific locations inside a virtual environment.

About 300 square feet of the overall game environment are devoted to the LAND category in Sandbox in Decentraland, the land pieces of a different size, measuring 50 square feet each.

Users who have amassed a sufficient number of parcels of land can collude their holdings into a single estate if they choose. One real-world illustration of this concept is "The Secrets of Satoshi's Tea Garden," a property on Decentraland that consists of 64 individual parcels of land. Because of its size and location, it fetched a price of 1.3 million MANA (approximately $80,000) when it was sold in 2019. Furthermore, the "land" is encircled by digital highways, which makes it relatively easy to get to.

The Metaverse Group spent 618,000 MANA, comparable to $3.2 million in 2021, to purchase a property in Decentraland in 2021. The transaction took place. Regarding digital real estate, location is just as important as the physical world. Plots located close to access points or sites like virtual arenas that are certain to have a high volume of virtual foot traffic typically have a higher value.

Rental services and money lending

People can make passive income by renting their NFTs out, although this option is contingent on the market demand. This is possible because of PARSIQ's IQ Protocol, a decentralised finance (DeFi) platform that offers game developers many revenue-generating opportunities. Landowners in the Metaverse can take advantage of this.

Through specified criteria negotiated with renters and enforced by smart contracts, the IQ Protocol enables virtual landowners to collect yield and rent fees analogous to traditional property and real estate dynamics.

Residual dividends

In addition to the royalties that can be earned by content creators when NFTs are sold or resold on secondary markets, passive dividends can also be earned by investors through the use of NFTs. One illustration of this is a portion of the virtual Monaco racing track that appears in the F1 Delta Time game and was sold at auction in December 2020 for $222,000.

The NFT that represents the digital track entitles the owner to collect 5% dividends from all of the events that take place on it. These dividends include yields from "Elite Events" that require participants to stake REVV to participate.

Token-based transactions and gameplay in the Metaverse

Earning in-game currency and other prizes through gameplay

The play-to-earn (P2E) paradigm has been implemented in games such as Axie Infinity and Aavegotchi to build new virtual economies. These new economies reward users with assets such as non-fungible tokens (NFTs) and in-game cryptocurrencies, which may be traded, sold, or borrowed. In addition, other games such as Battle Racer contribute to the expansion of utility by offering individual non-fictional items modelled like automotive parts. After that, users can either purchase these components to use in the construction of their own vehicles or sell them individually on secondary markets such as OpenSea.

The methods in which users are now utilizing NFTs in this blockchain networking can also be used for cosmetics and aesthetics in more mainstream games like Fortnite, but this implementation is still a few years away. This would not only assist players in more accurately reflecting the ownership and originality of their goods, but it would also raise the worldwide in-game expenditure, which is expected to exceed $74.4 billion by 2025.

Epic Games, the company that developed and owns Fortnite, will not actively distribute or engage with non-fictional items (NFTs) due to the prevalence of what it views as scams in the market. However, the company will gladly accept any game in its store that allows NFTs.

Avatars as well as social gatherings

Individuals collect NFTs for the social credit, status, and sense of belonging that can be found in a community of those who support the same causes as them, as seen by Twitter's recent acceptance of NFT profile images. In addition, people collect NFTs because they are entertaining.

For instance, one illustration of how avatars are developing in the Metaverse can be seen in the collaboration between Gutter Cat Gang, a collection of non-fictional characters, and House of Kibaa, an up-and-coming virtual reality studio. This partnership is an example of how avatars are evolving in the Metaverse. The community of House of Kibaa is eligible to participate in exclusive raffles for the chance to win free upgrades such as weapons, wearables, vehicles, pets, and real estate such as trap houses and mansions within the House of Kibaa Metaverse. In addition, House of Kibaa is working together to create animated 3D avatars available for all Gutter species.

Exclusive gatherings and celebrations

While this is happening, virtual events, parties, and concerts conducted in various Metaverses can make cash by selling NFT passes to attendees. For example, snoop Dog, an enthusiastic supporter of the NFT who is also a rapper and businessman, collaborated with The Sandbox to throw a private party in September of 2021. According to the announcement on The Sandbox, a total of 1,000 NFT party permits were distributed, of which 650 were made available for purchase on

The Sandbox's market. People who purchased The Snoop Private Party Pass were essentially granted access to Snoop Dogg's private lifestyle, as well as to unique NFTs, events, and the opportunity to have Snoop Dogg perform a private concert on their LAND.

To the next level: Sophisticated (NFT) trading in the Metaverse

According to estimates, total spending on NFTs would exceed $12.6 billion by the end of 2021. As a result, buyers are maximising the potential of their investments and collecting passive yields (also known as residual income) in several different methods, including the following:

Yield-generating NFTs:

Even for the most experienced investor, timing the markets on Web 3 can be next to impossible. As a result, to generate more practical incentives and mitigate the volatility of the NFT sector, many projects are generating passive returns by issuing governance tokens to holders. Among these collections, the Genesis Cyber Kongz is particularly noteworthy because it is programmed to create ten $BANANA tokens every day for the next ten years. There are also things like SupDucks ($VOLT) and Mutant Cats ($FISH) as more examples. Even though the use case is not entirely evident at this time, it was anticipated that Bored Ape Yacht Club would introduce its own coin in the first quarter of 2022.

In addition to collecting NFTs that produce passive yields, investors can also reap the benefits of NFTs and DeFi protocols by staking, which involves locking up their assets in a smart contract to receive rewards. This allows savvy investors to reap the benefits of NFTs and DeFi protocols simultaneously. Because of the stringent restrictions surrounding securities, the vast majority of these awards are in the form of unique tokens to the project. These tokens can be used for conventional governance or voting rights functions. On the other hand, stakeholders can liquidate their holdings in the market if the tokens are traded on decentralised exchanges such as SushiSwap or Uniswap.

In the past, NFTs were generally considered to be static constructs. For the most part, what someone minted or acquired on the secondary market could not be modified. However, people can now stack non-fungible tokens over other non-fungible tokens and transform a single non-fungible token into a virtual basket that can carry several ERC-based tokens, thanks to platforms such as Charged Particles. This new type of nested non-fungible token helps add intrinsic value to a collectable and transforms what would be a highly speculative investment into something resembling a yield-generating asset. It does this by attaching a scarce non-fungible token to another scarce non-fungible token, and so on, ad infinitum.

In general, layered NFTs provide value for users within the Metaverse by enabling users to customise virtual things and by introducing additional functionality to those products. During a webinar, Ben

Lakoff, co-founder and business lead of Charged Particles, presented a few examples of possible applications. They consist of a picture that can alter depending on the number of tokens deposited in it and a weapon that increases in power as the contained NFTs accumulate interest over time.

Strategies for generating residual income from non-financial assets

Rent out NFTs

Renting out your non-financial assets, particularly those in great demand, is one way to generate passive income for yourself.

For instance, some card trading games allow players to borrow non-fantasy trading cards (NFT) to increase their winning odds. To nobody's surprise, the transaction conditions that will be carried out between the two participating parties will be governed by smart contracts. Users of NFTs consequently have the ability, in most cases, to choose the length of the leasing agreement that best suits their needs and the lease fee for the NFT.

reNFT is an excellent illustration of a platform that allows users to rent or borrow NFTs. This gives lenders the ability to determine maximum borrowing durations and daily interest rates, which currently vary on average from 0.002 to 2 wrapped Ethereum (WETH). The native coin of Ethereum is called ether, and WETH is the ERC-20 version of ether (ETH).

Royalties from the NFT

Regarding the secondary market, the underlying technology that powers NFTs allows authors to specify terms that impose royalty costs anytime their NFTs change hands. In other words, even after selling their own creations to collectors, the artists still have the opportunity to receive revenue from passive sources.

Because of this, they have the chance to earn a portion of the selling price of the NFTs that are under discussion indefinitely. In addition, when an original piece of digital art is transferred to a new owner, the original creator is entitled to a royalty of ten per cent of the total sale price. This occurs whenever the original artist's work is resold.

It is crucial to remember that NFT developers often decide on these fixed percentages before minting the NFTs. In addition, the entire procedure for distributing royalties is governed by smart contracts, which are computer programs that can carry out their own execution and are used to enforce commercial agreements. Because of this, as a creator, you will not have to manually track payments or enforce the terms of your royalty agreements because the process is entirely automated.

Stake NFTs

The ability to stake non-fungible tokens (NFTs) is one of the advantages that come from the combination of NFTs and protocols for decentralized finance (DeFi). Staking is the practice of placing digital assets into a DeFi protocol smart contract to earn a yield. This can also be referred to as "locking away" digital assets.

Although some platforms accept a wide variety of NFTs, others require you to acquire native NFTs to earn staking token incentives (typically priced in the platform's native utility token).

The following are some examples of platforms that support NFT staking:

- Network de la Kira

- NFTX

- Splinterlands

- Only1

A portion of the benefits given out to stakeholders may sometimes be denominated in governance tokens. These types of protocols give the holders of these tokens the ability to vote on how their ecosystems will evolve in the future. It is feasible to reinvest coins generated through staking NFTs into other yield-generating protocols the majority of the time; however, this is not always the case.

Offer liquidity to earn NFTs.

Due to the ongoing integration of NFTs and DeFi infrastructures, it is now feasible to create your position in a specific liquidity pool by providing liquidity and receiving NFTs in return. This is done by providing liquidity to other participants in the pool.

For instance, when you offer liquidity on Uniswap V3, the automated market maker (AMM) will issue an ERC-721 token. This token, also known as an LP-NFT, contains information about your portion of the total amount locked away in the pool. The token pair you placed, the symbols of the tokens, and the pool address are all etched into the NFT as additional information. You can sell this NFT to liquidate your stake in liquidity pools rapidly.

Adopt NFT-powered yield farming as your primary method

Users can now farm for yields utilizing goods that NFTs power because NFTs are rapidly becoming an essential component of AMMs. Using several DeFi protocols to achieve the maximum potential yield with the digital assets you already possess is referred to as "yield farming."

Suppose we continue with the illustration from earlier. In that case, the LP-NFT tokens issued on Uniswap in the capacity of liquidity provider tokens can also be used as collateral or staked on other protocols to earn extra yields. Imagine this as receiving a yield in addition to the one generated by the other protocol. This potential opens the door to a multi-tiered approach that generates cash and is perfect for yield farmers.

It goes without saying that it is extremely important for you to keep in mind that NFTs and the technology underlying smart contracts are still in their infancy. As a consequence of this, the majority of the applications that can provide the potential discussed in this book are still in the process of being developed. Therefore, before deciding to use any of the tactics outlined above, it is crucial to conduct research as required and understand the dangers associated with doing so.

NFTs should be used to implement a Metaverse

You may have noted that the Metaverse is still a relatively new concept, and only a small handful of businesses have already produced practical solutions in this field that incorporate NFTs. This may be because the Metaverse is still quite difficult to build. Therefore, if you have a use case for the combination in your company and the means to make it happen, you may be one of the first organizations in your industry to capitalize on these two trends. This is oh so true if you have a use case for the combination in your business.

Because the majority of businesses do not currently have any VR developers working for them on a full-time basis, we could suggest trying to collaborate with a company that has many years of experience in the field of developing immersive application software. Even if a company has a few local developers, those developers might not have platform-specific understanding of VR software, such as how to create using Unity and Unreal or how to incorporate movement tracking. In this case, the company might want to think about going this route anyhow. In the event that you do not intend to handle the technical components of your endeavor on your own, having knowledge of blockchain technology and NFT minting will be of assistance.

I hope that you now understand the basic ideas included inside the two concepts and have a general grasp of the direction in which the market is moving. If you want to differentiate your company from others in its industry and present yourself as one that is focused on the future, embracing the Metaverse and NFTs could provide you with an excellent opportunity to do so.

Chapter 8

Metaverse vs. Virtual Reality

A significant number of works of science fiction have given us glimpses of virtual worlds in which individuals are free to take any identity they so desire. One of the most intriguing aspects of today's technology is the idea of creating a virtual world in which one can live without having to deal with the challenges of the physical world. You have probably been hearing quite a lot about the latest happenings in the Metaverse lately.

On the other hand, because of the long-standing connection between the Metaverse and VR technology, there has been a lot of buzz surrounding similarities between the Metaverse and virtual reality.

People unfamiliar with the idea might think that the Metaverse is essentially a virtual environment that can be accessed using virtual reality headsets. Is there a difference between Metaverse and virtual reality? No. A lot of information available can help you differentiate between the Metaverse and virtual reality. A comprehensive look into both Metaverse and virtual reality will be provided in the following discussion, which will assist you in determining the distinction between the two technologies.

Excitement about virtual reality and the Metaverse

After major corporations made a significant financial investment in the Metaverse, it quickly became possibly the most popular topic of conversation in the technology sector. Many of the world's most influential companies, including Facebook, Microsoft, Google, Sony, and dozens of others, are moving their platforms into the Metaverse. After the revelation that Facebook would change its name to Meta in October 2021, the idea of a Metaverse began to attract widespread attention. What prompted Facebook to make such a significant move?

Comparisons between virtual reality and the Metaverse have garnered much attention recently, bringing up an interesting point. The business recently announced that it had reached a milestone by selling 10 million Oculus VR headsets. In most explanations, the Metaverse is portrayed as being analogous to the practice of using the Internet while present in a virtual environment. Virtual reality, on the other hand, would play a supplementary role in engaging with the Metaverse. Therefore, let's learn more about virtual reality and the Metaverse to determine the most effective technique to discern between the two.

Shouldn't Virtual Worlds Already Be Considered Part of the Metaverse?

When you consider the potential distinctions between the Metaverse and virtual reality, you are bound to entertain these uncertainties in your mind. The Metaverse is not simply any of the several virtual worlds that can be accessed through VR technology. On the other hand, it will be a collaborative and ever-present virtual environment that will supplement the real world. Some of the most important answers to that question can be found by delving more deeply into how the idea of the Metaverse is distinct from the Internet as it exists now.

You would not simply be a device in the Metaverse that requests information from web servers. Instead, users of the Metaverse can access online material and virtual environments in the guise of a person using a digital avatar. Users can participate in various activities within the Metaverse by controlling a virtual avatar that inhabits the simulated environment. In reality, the Metaverse would function as a vast, cooperative online environment in which users would not be subject to any limitations. For instance, it is unnecessary to log out of a game to go to a virtual museum when you are in the Metaverse.

What exactly does "Virtual Reality" mean?

The term "virtual reality" (VR) refers, as its name suggests, to the development of artificial worlds facilitated by technological means. A computer-generated three-dimensional world is meant by the term "virtual reality" when discussing its application in technology. An individual can explore the virtual reality experience and engage with it in a setting that is both immersive and engaging.

Users are entirely submerged in the worlds and can move around, interact with, and complete various tasks within virtual reality settings. So, it may come as no real surprise to you that the definition of virtual reality (VR) obscures, to some extent, the distinction between the

Metaverse and virtual reality. It is essential to understand that virtual reality can only provide access to confined experiences or regulated virtual worlds.

The use of specialized VR headsets and glasses, coupled with the computer system executing the programs allows a person to enter particular VR experiences. Furthermore, the specialized hardware contributes to an increase in the level of immersion that users experience with the sensory components, including sight and sound, present in the virtual environment.

You can also find specialized VR systems that let users don gloves with electronic sensors. These are also available for purchase. At this point, the entertainment industry has been shown to have the most significant uses for virtual reality. On the other hand, recent years have seen a rise in the popularity of utilizing virtual reality for educational purposes and in the fields of medicine and the armed forces.

The Difference Between VR and the Metaverse

Virtual reality vs Metaverse comparison can be defined using the fundamental outline provided by the detailed review of both virtual reality and the Metaverse. Virtual reality is almost always mentioned at some point during conversations pertaining to the Metaverse. As such, it really can be quite a challenge to get around the parallels between the Metaverse and virtual reality. On the other hand, the two are not even somewhat similar. The following is a list of the primary distinctions between the Metaverse and virtual reality:

Ownership

The next major point of differentiation between virtual reality and Metaverse would presumably allude to the different ownership options available for each. When you use a virtual reality (VR) system, you are essentially having an experience with a system that a brand owns. The brand is the proprietor of the content that can be found within the virtual reality experiences offered by the system. The only thing that you have under your control with virtual reality technology is the equipment.

Conversely, the Metaverse is an entirely new ballgame since it allows people to acquire ownership of virtual assets and experiences. It is irrelevant whether it's a piece of virtual real estate or an artefact - everything you produce and own in the Metaverse is your property. Users are granted the rights and responsibilities associated with full ownership when using the Metaverse.

Technologies

In the ongoing discussion between the Metaverse and virtual reality, bringing attention to the constraints imposed by technology would be an additional key highlight. When you examine virtual reality, you will notice that it has a few restrictions. Only as advances are made to VR technology will it be possible to produce increasingly realistic virtual experiences. In the end, virtual reality would be nothing more than playing out simulations and experiencing exciting virtual adventures.

Metaverse does not have any restrictions of this kind. The Metaverse does not solely derive its power from the technologies behind virtual reality. The features of the Metaverse are driven by a wide variety of technologies, including augmented reality, blockchain, cryptocurrency, and connection technologies. The Metaverse is a large virtual environment that permits the integration of new technologies to provide increased functionality. This is maybe the single most essential aspect of the Metaverse.

Experiences

When comparing VR with the Metaverse, the method in which the user experiences each is a crucial pointer that should be considered. In the case of VR, you will get the impression that you are wearing a headset or another piece of equipment that enables you to experience the virtual setting. On top of that, the experiences that may be had in virtual reality are restricted to a predetermined number of people, such as those participating in the game.

By integrating augmented and virtual reality technologies, the Metaverse creates a virtual cosmos that is comparable to the actual world. The result is that you will feel as though you are travelling around the real world while experiencing the virtual realms, albeit in the form of a digital avatar. The thing to note is that the experiences that can be had in the Metaverse are not confined to any particular location. The ability of participants in the Metaverse to visit, experience, and connect with other users and other areas in the Metaverse is one factor that contributes to the qualitative differences between the Metaverse and virtual reality.

Persistence

The concept of persistence is the following indicator that you can use to differentiate between the Metaverse and virtual reality. When contrasting virtual reality with the Metaverse, it is clear that virtual reality, sometimes known as VR systems, is one of the frontrunners. While you now have technologies such as VR that allow you to experience virtual worlds, the Metaverse is still being developed. However, once you turn off a virtual reality (VR) system, your experiences while using it are no longer available to you.

Virtual reality (VR) is turned on when it comes to persistence thanks to the Metaverse, a shared and persistent universe. So even if you leave the Metaverse, your digital avatar will remain. This is true even if you log out of the Metaverse. It would function normally even if other people in the Metaverse communicated and engaged with it.

The **concluding** analysis of the distinctions between the Metaverse and virtual reality demonstrates that the Metaverse encompasses more than just VR. The Metaverse development requires virtual reality as one of the cornerstone technologies. Users are granted access to the Metaverse as a result. However, virtual reality alone can only accomplish a limited number of tasks.

On the other hand, the Metaverse is a vast virtual universe that is still expanding and eventually will incorporate a three-dimensional depiction of the Internet. Users can navigate various virtual 3D places in the Metaverse as they would navigate through different web pages. The essential thing to take away from this is that the opportunities for integrating innovation into the Metaverse are virtually limitless. Begin your exploration of the Metaverse and determine how virtual reality can enhance your experience of it.

Chapter 9

How to Create a Metaverse Avatar

What is an avatar?

A person's virtual representation of themselves in a video game, social media site, etc., is known as an avatar. Consider an avatar representing you in virtual reality when the technology is available (Virtual Reality). According to technical specialists, an avatar is equally crucial when discussing the Metaverse. Why? Considering that it is how you represent yourself online.

Consider it the same way you would create a social network account. Typically, you will post a photo of yourself to the network to serve as a visual representation of who you are. Similar circumstances apply when dealing with the Metaverse; an avatar is required. However, since you choose a figure that you may dress and customize to your preferences, you don't need a genuine photo in this situation.

What Does the Metaverse Mean by 'Avatar'?

As was already noted, the Metaverse is a technological development that combines numerous platforms, like social media and video games. As a result, it makes it a whole lot easier for individuals to communicate with one another. According to reports, researchers frequently link the Metaverse and avatars together. They exhibit a dependent relationship, in other words. However, comparing the two, you probably don't realize why avatars are crucial in the Metaverse. And that's the topic this part wants to cover.

What role do avatars play in the Metaverse community? When interacting with the digital world, distinct people's representations are referred to as avatars. Simply put, you represent yourself in the virtual world using an avatar. For example, Avatar Metaverse gives users a sense of identity while using the Internet. You communicate your identity to the online community via an avatar.

However, it should be noted that interoperability—the capacity for avatars to communicate between different platforms readily—is a crucial aspect of avatars and is why it is so vital in the Metaverse. For instance, this capability enables you to move between several web platforms when you make purchases for your Avatar. In the Metaverse, avatars promote a feeling of community.

How to Easily Create a Metaverse Avatar

If you've heard of Metaverse avatars, you presumably already have an idea of what you envision the virtual world used to identify you. But because the Metaverse technology is so new, not many individuals know what to do. You may not get all the correct answers in this guide. We showcase the best Avatar creation tricks to enhance and expand your Avatar Metaverse experience. Do a little research to expand your knowledge.

You will currently come across quite a selection of Metaverse avatar builders, including the ready player me Avatar. This choice is trendy because you may customize it to look like you. First, however, some fundamental guidelines must be followed, and this section will concentrate on avatar construction.

Find a top-notch avatar creation program

There are numerous programs available that enable you to create an avatar. Choose the best tool for your needs based on the ecosystem you are using; for instance, users of Android and iOS might choose the Zmoji app.

Open the application to begin creating your Avatar

Most software you encounter will ask about your gender to determine the best style for you. You have the option to add a photo so that the avatar generator can create a character that closely resembles you. Additionally, you have the opportunity to settle for one of the ready-made players, my avatars.

Make the figure unique

You can choose from various facial traits in this step, including the size and shape of your nose, lips, eyes, hair, and eyes. You can alter your Avatar to reflect your ideal self.

Choose the most acceptable option, and then save your Avatar so you may share or use it in many digital settings. One of the easiest, and by far the fastest ways to construct an Avatar Metaverse that is ready for use after the phases mentioned above describes the development phase.

Remember to bring your NFT avatar creators to the Metaverse

You may infer from Metaverse's definition that computer scientists are working to build a reliable online community where people can engage freely. Every user of these digital networks also has a unique avatar. As a result, the community is open to all applications, including those from NFT creators of Metaverse avatars.

Today, most users accept the Metaverse avatar NFT that developers produce using unique algorithms. Because of their outstanding features, their popularity has continued to grow over time, and the businesses that created these Metaverse avatars NFT have sold out millions of units.

You've probably heard of the Filmora program if you enjoy editing videos. For editing videos, it is a widely used application. However, since the advent of Metaverse avatars, the software can now produce avatars that are ready for me to use. After all, the program includes stickers for augmented reality. When creating NFT Metaverse avatars, AR stickers are frequently helpful. Wondershare Filmora is the best option if you want to design your NFT avatar. So, use Wondershare Filmora to continue creating Metaverse avatars.

We frequently hear about the Metaverse world from technical specialists now that this technological development has completely transformed the Internet. Furthermore, to claim that we are well-informed about this expanding group would be a deception. After all, it symbolizes the development of the Internet, yet nothing is known about it. We're sure that Metaverse Avatars are necessary for its success, though. In light of this, this page focuses on outlining information regarding avatars and demonstrating how to create one. See the points above to learn more about avatars in the Metaverse.

Chapter 10

All You Need to Know About Web 3.0

Web 3.0 is predicted to be the following:

- **Open** – Content platforms will be created using open-source software.

- The network will be protected to the edge, and everyone will use zero trust.

- **Distributed** – Without the consent of a centralized authority, communication between devices, users, and services will be possible.

In the next phase of the Internet, people will be able to communicate directly, thanks to blockchain technology. Users will interact by joining a Decentralized Autonomous Organization (DAO), an association controlled and owned by its members. In addition, the user's data will be safeguarded through a network of publicly accessible smart contracts. These contracts will be kept on a blockchain decentralized network, which nodes will manage.

Here are some further Web 3 predictions:

Data transfers will be decentralized, and all transactions will be recorded on a distributed ledger that employs blockchain technology. Furthermore, thanks to open-source smart contracts, people won't have to rely on a centralized institution (like a bank) to preserve data integrity. As a result, the Metaverse will considerably boost the entertainment industry's revenue.

Consumers can instantaneously generate digital goods and non-fungible tokens (NFTs), which will safeguard intellectual property and personally identifiable information, thanks to blockchain technology (PII). It will be possible to make money off of user data.

Web 3.0

Web 3.0 was initially referred to as the Semantic Web by Tim Berners-Lee, a programmer who developed the World Wide Web (WWW). Berners-Lee imagined an intelligent, independent, and open Internet that used AI and machine learning to act as a "global brain" and interpret content conceptually and contextually. Unfortunately, this idealized version didn't exactly work out due to technical limitations, such as how expensive and difficult it is to translate human language into machine-readable language.

Here is a list of typical Web 3.0 characteristics:

- Through search and analysis, the semantic web, a development in online technology, enables individuals to create, share, and connect content. It focuses on word understanding rather than employing keywords and statistics.

- Artificial intelligence and machine learning are used. As a result, Web 3.0 was created to become more intelligent and more reactive to consumer requests. The result is a computer that employs

Natural Language Processing (NLP) if these concepts are combined with NLP.

- It demonstrates how the Internet of Things (IoT) connects numerous devices and programs. Semantic metadata enables this process to utilize all available data effectively. Without a computer or other smart device, anyone can access the Internet anytime and anywhere.

- It provides "trustless" data and enables users the option of interacting in public or privately without putting them in danger by using a third party.

- Their 3D graphics are used. E-commerce, virtual tours, and computer games have already made this clear.

- It facilitates involvement without requiring approval from a central authority. Without permission, it is.

It can be used for:

- Metaverses: An infinite, virtual world with 3D graphics

- Blockchain video games support the NFTs' principles by letting players own in-game assets.

- Digital infrastructure and privacy: This application uses more secure personal data and zero-knowledge proofs.

- A decentralization of finance. Examples of this use include peer-to-peer digital financial transactions, smart contracts, and cryptocurrencies.

- Decentralized autonomous organizations. The members of online communities own those communities.

Web 2.0

Web 2.0 consists of many people providing even more content for an expanding audience, whereas Web 1.0 consists of a small group of people producing information for a larger audience. As a result, web 2.0 emphasizes involvement and contribution more than Web 1.0 focuses on reading.

The primary goals of this Internet form are User-Generated Content (UGC), usability, engagement, and increased communication with other systems and devices. The user's experience is crucial in Web 2.0. This Web form was in charge of creating social media, teamwork, and communities as a result. Therefore, most users in today's world see Web 2.0 as the dominant mode of web interaction. Web 1.0 was often referred to as "the read-only Web," but Web 2.0 is referred to as "the participative social Web." Web 2.0 is an enhanced and expanded version of Web 1.0, thanks to the addition of web browser technologies like JavaScript frameworks.

The following is a breakdown of Web 2.0's typical characteristics:

- It has interactive elements that respond to user input.

- Application programming interfaces are used (API)

- It promotes self-use and permits involvement through podcasting, social media, tagging, blogging, comments, RSS curation, social networking, and web content voting.

- Users can retrieve and classify data collectively using its free information sorting service.

- It makes use of newly created application programming interfaces (API)

- It employs information that has been developed; society as a whole uses it, not just particular communities.

Web 3.0's Function

Your data is kept in Web3 on your cryptocurrency notecase. Through your wallet, you will connect with apps and communities on web3, and you may take your data with you when you log out. Since you are the data owner, you may hypothetically choose whether to monetize it. Now that we have our guiding principles, we can consider how specific web3 development features are intended to achieve these goals.

When you use a website like Facebook or YouTube, these companies collect, own, and recoup your data. Your information is kept on your web3 cryptocurrency wallet. Through your wallet, you will connect with apps and communities on web3, and you may take your data with you when you log out. Since you are the source of the property, you may theoretically choose whether to commercialize it.

Pseudonymity:

Like data ownership, privacy is a characteristic of your wallet. On web3, your wallet acts as your identification, making it challenging to link it to your real identity. So, even if someone watched a wallet's behavior, they wouldn't be able to tell which wallet it was. "My private information is concealed, but my actions are obvious." Neuroth used it as a quote. Customers can connect to their crypto wallets utilized for illicit activity using services. Your identity is hidden for everyday use, nevertheless.

Although wallets boost the amount of privacy for Bitcoins, privacy currencies like Zcash and Monero completely conceal the identity of both the sender and receiver. Observers can follow transactions on blockchains for privacy coins but cannot see the wallets involved.

Decentralized autonomous entities that execute apps will be present in Web3 (DAOs). As a result, users who hold governance tokens are now made decisions instead of by a centralized authority. The tokens can be obtained by participating in the upkeep of these decentralized programs or by purchasing them.

The CEO of a typical firm oversees the carrying out of changes that the shareholders have accepted. A DAO's token owners can vote on changes, and a smart contract immediately incorporates them into the DAO's code if they are accepted. Since DAOs are democratized, everyone has access to their source code.

Principal Uses of Web 3.0

Web 3.0, which has blockchain at its foundation, enables a wide range of new apps and services, including the ones listed below:

- **Non-fungible Token**s (NFTs) are tokens held in a blockchain with a cryptographic hash and are individually unique.

- **Decentralized finance** (DeFi) is a novel use case for Web 3.0 that enables the provision of financial services outside the bounds of traditional centralized banking infrastructure. Decentralized blockchain technology is being used as the foundation for DeFi.

- **Cryptocurrency:** Through Web 3.0 applications like cryptocurrencies such as Bitcoin, a new monetary universe, distinct from the conventional world of fiat cash, is being formed.

- **Decentralized applications** (dApps) are computer programs that are logged in an immutable ledger and are run programmatically. They are constructed on top of the blockchain and employ smart contracts to speed up the delivery of services.

- **Bridges that cross chains**: In the Web 3.0 era, there are many different blockchains, and bridges that cross chains offer some connectivity between them.

- **DAOs:** With their ability to provide some structure and decentralized governance, DAOs are prepared to fill the position of Web 3.0's governing bodies potentially.

The Future of The Internet

The world is moving toward an Internet where users have total control over their data and privacy while allowing businesses to use it (or not). It will be feasible to do all of this thanks to blockchain technology. Web 3.0 will speed up user data's fair and transparent use, enabling 3D visuals, cross-platform development tools, and tailored search results. In the coming years, the Internet will get more immersive and exciting.

Chapter 11

Metaverse: Top Benefits and Drawbacks

There are some intriguing benefits to working in the Metaverse. As the epidemic has demonstrated, organizations can succeed with alternate working arrangements, from full-time remote to hybrid working settings. The pandemic has also shown that people prefer to work in this fashion and benefit from the advantages it has on their general mental health, work/life balance, and well-being by not having to endure a long drive each day to and from the workplace.

As an illustration, many people claim to have relied on a daily nap during the global epidemic. Some offices now offer napping areas to keep this routine for their employees. Employees' autonomy over their schedules and procedures during the day has enhanced productivity and job satisfaction. Theoretically, working remotely while feeling present—or at least your avatar present—in a workspace where you can interact with co-workers and manipulate products and plans for work projects. With the Metaverse, you get the sense that a live, interactive, cohesive work experience without leaving your home is possible.

Since its inception, Internet technology has increased our ability to choose where and how we work. The pandemic presented the watershed moment for a true cultural shift in how we think about work beyond the usual 9–5 office mentality, even if the possibility to work remotely has been for decades. And the Metaverse offers a plethora of solutions to lessen the difficulties that remote work may provide.

There are many instantly apparent benefits of working in the Metaverse that goes beyond its sci-fi 'wow' aspect and offers advantages on a practical level. The following are some benefits of operating in the Metaverse:

Improving the efficiency of remote work

Working face-to-face gives you a better understanding of how clients and team members feel, primarily through body language. The effectiveness of communication may suffer from the lack of face-to-face engagement. The Metaverse, however, allows for more interaction in a 3D environment where employees can congregate much as they normally would in their "real world," going beyond traditional telecommuting as we know it. Team productivity is quantified and enhanced by this level of visibility and connectivity.

Using 3D visualization, address issues

Visual problem solving is essential for ensuring success in many areas. These range from healthcare to building, architecture to life sciences. 3D visual modeling offers a distinct, precise, and efficient way to develop, solve problems, collaborate, and create in the Metaverse. For instance, a group of architects can produce mock-ups that correspond to actual specifications and base judgments on these models.

Gain access to interoperability and unlimited space

Although the Metaverse resembles the real world, it has the benefit of being able to grow indefinitely. You can add as much area and amenities as you need to manage your business, from whiteboards to conference room space. Without spending money on infrastructure, you may expand further.

Eliminate reliance on hardware

In the Metaverse, employees—or their 3D avatars—can meet face-to-face without additional hardware, making conferencing equipment obsolete. Digital whiteboards and workstations will replace the current requirement for such equipment expenditure.

Problems with operating in the Metaverse

The potential effects of this digital work experience in the Metaverse worry many sceptics. One problem is privacy.

Hardware costs

The cumbersome and expensive nature of the current technology required to realize the full potential of the Metaverse is another potential drawback, or at the very least, a challenge that must be addressed. The issue of improving and making these tools accessible—and appealing—is one that Bill Gates is prioritizing.

In last year's blog, Gates stated, "The concept is that you will eventually employ your avatar to congregate with others in a virtual place that simulates the sense of being in an actual room with them.

"To do this, you'll need motion capture gloves and VR goggles that can precisely capture your facial expressions, body language, and voice quality. Because most individuals don't yet own these tools, adoption will be somewhat slower. (The fact that many individuals already had PCs or phones with cameras was one of the factors that made the switch to video meetings possible so quickly.) Microsoft will release an interim version that uses your webcam to animate an avatar used in the current 2D setup early next year.

Businesses that have adopted the Metaverse

Microsoft and Facebook are the leading brands you will see advancing and supporting the Metaverse (Meta). To support its Metaverse vision, Meta has invested $10 billion in creating and acquiring hardware and software, including VR and AR technologies.

Microsoft stated earlier in 2022 that it would buy video game publisher and developer Activision Blizzard as a first step towards staking a claim in the Metaverse. It is clear from this $70 billion deal, Microsoft's most significant yet, that it will wager heavily on the Metaverse in the coming years. Sundar Pichai, the CEO of Google, also emphasizes his company's interest in AI, despite the relatively disastrous 2014 Google Glasses project.

But not only tech industries are affected. Shopify wants to rule the augmented reality market in terms of online purchasing. By selling digital clothing modelled after the actual Balenciaga collection in a virtual store in the video game Fortnite in 2021, the fashion label Balenciaga entered the Metaverse. Events are planned, and viewing experiences are improved by musicians, museums, and the sports business. With multimillion-dollar sales, the blockchain projects Decentraland and The Sandbox launched the virtual real estate market.

Metaverse doubters

Despite the hoopla around the Metaverse, not everyone is convinced that building a virtual world that coexists with the actual one is what people eventually desire. For example, in a 2021 interview with CNBC, Strauss Zelnick, CEO of gaming development company Take-Two, stated his scepticism. He said, "I'm sceptical that we're going to wake up in the morning and consciously sit at home, strap on our headsets, and conduct all our daily activities that way. We didn't particularly enjoy having to do that during the pandemic.

It is inevitable that the tools and technology that make up the Metaverse vision will significantly impact how, when, and with whom we work because there is a wealth of data that supports the advantages we experienced when more flexible working arrangements were not only possible but required. However, remembering to live as fully in the actual world as we can in the Metaverse may yet prove to be the most formidable obstacle.

Conclusion

Non-fungible tokens, Metaverse cryptocurrencies projects, etc., are hopefully well-explained so that even beginners can understand the basics. The level to which governments, public and private sectors, individuals, developers, and investors work on Metaverse shows a glimmer of hope. Individuals attempting to enter the Metaverse should know its fullness, as it can be volatile.

What investments are real if the Metaverse is a fantasy? It is a unique technology that the world must understand and adopt. Every fantastic technology has flaws. So, its success depends on the parties involved. Several attempts to launch Metaverse into the world have shown a considerable increase. Facebook sponsors Metaverse.

As described above, it changed its parent business name to Meta to show its dedication to the Metaverse concept. This new technology is risky. However, its contributions to health care, education, manufacturing, and fashion have improved lives and the economy.

Blockchain is a decentralized mechanism that would help the Metaverse. Without law and order, chaos and abuse are inevitable. Metaverse encourages inventiveness and infinite experiences. The Metaverse helps us navigate the unknown.

Web3 helps extend and interoperate Metaverse. True flexibility requires interoperability; few people can multitask. Moreover, interoperability enables the Metaverse to merge apps and improvements. This interoperability is crucial for the Metaverse vision to handle a significant population from diverse continents.

As the Metaverse unfolds, we question how all this will be possible, yet the uptrends and success of Metaverse cryptocurrency initiatives remind us that it is possible. Armed with the information within this guide, you can join the Metaverse and maximize its potential. We will update this guide as things develop - which inevitably they will.

Thanks for reading